Nomads of the 19th Century Queensland Goldfields

Lennie Wallace

AF585166

Nomads of the 19th Century Queensland Goldfields

The Goldfields of Gympie, the Palmer and the Hodgkinson

including

The Life Story of Dr. Jack Hamilton 1842–1916
Prospector, Pugilist, Physician, Politician

Lennie Wallace

© Lennie Wallace 2000 - 2012

This book is copyright. Apart from any fair dealing for the purpose of private study, research, criticism or review, as permitted under the Copyright Act, no part may be reproduced by any process without written permission. Enquiries should be addressed to the Publisher.

All rights reserved.

First published in 2000 by Central Queensland University Press

Second published in 2012 by Boolarong Press, Salisbury, Brisbane, Australia.

National Library of Australia Cataloguing-in-Publication entry

Author:	Wallace, Lennie.
Title:	Nomads of the 19th century Queensland goldfields / Lennie Wallace.
ISBN:	9781921920592 (pbk.)
Subjects:	Hamilton, Jack, 1842-1916.
	Gold miners--Queensland--Biography.
	Politicians--Queensland--Biography.
	Gold miners--Queensland--History--19th century.
	Gold mines and mining--Queensland--History--19th century.
	Frontier and pioneer life--Queensland.
	Queensland--History.
	Australia--Gold discoveries.
Dewey Number:	622.3422092

Cover Design and Typesetting by Ampersand Design - Kurt Otto

Front cover illustration by J. A. Turner *The Precious Metal* 1894. oil on canvas 133 x 90.6 cm (image) sight. Collection: Dixson galleries, State Library of New South Wales. Reproduced with the permission of the State Library of New South Wales.

Printed and bound by Watson Ferguson & Company, Salisbury, Brisbane, Australia.

INTRODUCTION

The search for gold has always been part of the Australian culture. True, a convict's claim to have found a payable mine in late August 1788 proved false but gold was discovered by convicts making a road to Bathurst sixteen years later. At the time, a gold rush was the last thing the authorities wanted added to their already overfull plate and silence was kept on the find by 'menaces and flogging'.

Hargreaves, back from the Californian gold rush, found gold in creeks near Bathurst in early 1851. The government couldn't control it but decided it may as well profit from it. It introduced a Goldseeker's Licence of thirty shillings a month, a very considerable amount. Despite this impost, by the end of 1851, 25,000 miners were licensed in New South Wales. Ballarat followed on Bathurst's heels and nuggets weighing in the hundreds of ounces were found in both colonies.

There was an abortive rush to Canoona near Rockhampton the year preceding Queensland's separation (1858). It proved 'disastrous to the 15,000 – 20,000 adventurers who swarmed to the Rockhampton district in search of 'saint-seducing' gold.' Highlighting the fickle nature of gold prospecting, Mt. Morgan, the 'mountain of gold' was a mere 'dozen miles or so' from Canoona. And produced over fourteen and a half million pounds worth of gold. Peak Downs, inland from Canoona, produced payable gold in 1862.

Gold was found near Bowen in 1865 and also on the Star River inland from what was to become Townsville. However it wasn't until Nash discovered the Gympie goldfield in 1867 that the colony entered the upper ranks of gold producers. There were big nuggets found there also. One weighed almost one thousand ounces and was valued at 3,675 pounds. In 1869, Richard Daintree, photographer, pastoralist, explorer and geologist found gold at Gilbert River which led to the formation of the township of Gilberton. The following year, a crusher was processing ore at Ravenswood a major field west of the Dividing Range from Townsville. Silver was also found there and at Argentine further upstream on the Burdekin River.

In 1871 Jupiter Mosman, an aboriginal horseboy employed by Hugh Mosman, discovered gold at Charters Towers. This was worked by Mosman and his partners George Clark and Fraser but the news soon leaked out and miners flocked in turning Charters Towers into a very prosperous township. Maurice Fox's Durham mine at the Etheridge (Georgetown) was in production when Hann's surveyor Frederick Warner found gold in the Palmer River. The exploration party, considering the isolation of the field from any settlement and remembering the fiasco of Canoona, did not promote the discovery but news of the find did get out and it was left to the Prince of Prospectors, James Venture Mulligan and his partners to prospect a further twenty mile stretch of the Palmer River. What they found left no doubt but that it was a bonanza and the Palmer Gold Rush began. Lesser fields were abandoned as miners moved eagerly north to the Palmer.

Watson, Russell and Verge found gold near Coen in 1875. This was followed by gold strikes on the Hodgkinson (closer to Mareeba) and the Tate. As the easy gold on the Palmer was declining, miners left that field for the Hodgkinson and Mareeba district fields. Some prospectors headed further north and discoveries were made at Coco Creek, north of Cooktown by the Webbs in 1890, the year the Quetta sank with 133 passengers and crew. Cairns and Bowden found gold at Starcke also near Coco Creek. Two years later Willie Baird found gold at Bairdsville on the Batavia River north of Coen. In the years that followed other gold strikes were made in the far north. Klondyke Creek near Coen was one of the early ones.1900 saw the declaration of the Hamilton Gold Field discovered by John Dickie at Ebagoola between Cooktown and Coen.

Large quantities of gold proved elusive but small finds encouraged prospectors. Willie Baird's aboriginal helper, Pluto, discovered payable gold and gave his name to the settlement of Plutoville. (1911) Later again, an aboriginal woman called Kitty Pluto found gold on the Wenlock River. The legendary Big Strike in the furtherest north of the Peninsula is still lying in wait for a discoverer though small strikes have been made both of the mainland and on the adjoining islands.

Charles Arrowsmith Bernays in his book Queensland Politics During Sixty Years 1859-1919 gives the following figures for gold production to the year 1917:

- Gympie (1867) 3,308,156 ounces
- Charters Towers and Cape River (1875 and 1867) 6,622,261 ounces
- Croydon (1886) 766,996 ounces
- Etheridge, Oaks and Woolgar (1869,1908 and 1882) 605,797 ounces
- Hodgkinson (1875) 229,706 ounces
- Palmer (1873) 1,329,042 ounces
- Hamilton (1900) 45,547 ounces
- Coen (1893) 51,255 ounces

The total gold production for Queensland to the end of 1917 was 19,330,803 ounces with a value then of £82,111,979. Copper brought in a further £20 million, silver just under two million, coal seven and a half million, tin was worth over nine million and opals and other gems were mined to a value of half a million pounds. The total of mineral products mined for the period was £41,958,091.

ACKNOWLEDGEMENTS

The Cook Shire for help with R.A.D.F. assistance and personal encouragement. To John Oxley Library, Bowen Historical Society, D. Langmore and G. Bolton, Jan Wegner, Jean Hayden, Peter and Lorraine Ryle for much appreciated research material.

CONVERSION TABLE AS USED BY DEPARTMENT OF MINES

Length

1 foot	=	0.305 metre
1 yard	=	0.914 metre
1 mile	=	1.61 kms

Mass

1 ounce fine	=	0.031 kilograms
1 pound	=	0.453 kilograms
1 ton	=	1.02 tonnes

Conversion of pounds to dollars would be misleading if the equation one pound equals two dollars is used. There is little relationship to the pound of the late 19th century to two dollars of the late twentieth century. A skilled miner's wage for a 48 hour week in the period 1884-1914 varied from three pounds to five pounds per week

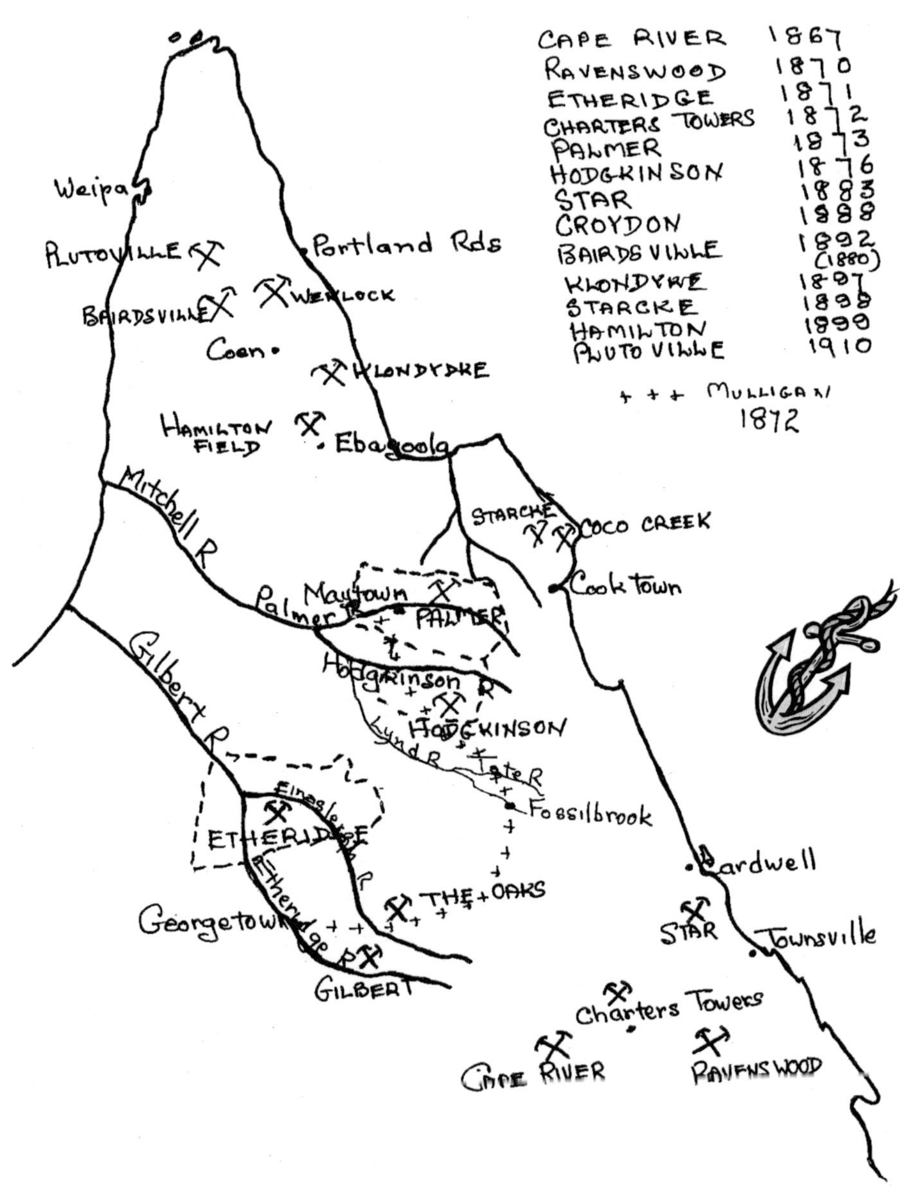

CAPE RIVER 1867
RAVENSWOOD 1870
ETHERIDGE 1871
CHARTERS TOWERS 1872
PALMER 1873
HODGKINSON 1876
STAR 1883
CROYDON 1888
BAIRDSVILLE 1892 (1880)
KLONDYKE 1897
STARCKE 1898
HAMILTON 1899
PLUTOVILLE 1910
+ + + MULLIGAN 1872
Weipa
PLUTOVILLE
Portland Rds
BAIRDSVILLE
WENLOCK
Coen
KLONDYKE
HAMILTON FIELD
Ebagoola
Mitchell R
STARCKE
COCO CREEK
Cooktown
Maytown
Palmer R
PALMER
Hodgkinson R
HODGKINSON
Lynd R
Tate R
Fossilbrook
Gilbert R
ETHERIDGE
Etheridge R
THE OAKS
Georgetown
GILBERT
Cardwell
STAR
Townsville
Charters Towers
CAPE RIVER
RAVENSWOOD

CONTENTS

6

Scandals & Squabbles. The Ninth Queensland Parliament Develops the Far North
1883

7

Capital, Unions, Patriotism and the Advent of the 1890s Depression

8
The Rise of Labour in Queensland Politics

9
Return of the Miner
1903-1911

10
The Death of the ANZACs and the Death of Noble Dr. Jack
1915-1916

To my mother Sadie Waddell nee McNutt and to Elwyn of the Hodgkinson who began our involvement by disapproving of 'Dr.' Jack but who ended up as his second greatest living fan.

1

THE RETURN OF THE NATIVE 1863

John Hamilton Dinwoodie returns to Australia as John Hamilton. Lands at Port Denison, December 1863. Decides against a career in pastoralism and joins rush to the goldfields. Meets Macrossan at the Ridgelands workings. Dalrymple and Jardine in Rockhampton. Crocodile Creek at Mt. Morgan.

DR JACK HAMILTON LANDS IN PORT DENISON

Ever since Jack Hamilton left Australia as a lad named John Hamilton Dinwoodie he had a persistent longing to return. There was no definite reason. The other members of his family were content back in Britain, but something deep within him insisted his future lay not in the Old Country but in the new. And of course, he was Australian by right of birth.

But when he arrived in Melbourne the city failed to realise the warm feelings of familiarity he'd hoped for. Though the wealth from the Victorian goldfields led to the establishment of a substantial town, Jack found it bore little resemblance to the small township in which he'd spent his childhood. There were no regrets in leaving, only an impatience to be gone when the steamer pulled out from the cold, grey waters of Port Phillip Bay. The exciting prospect of Queensland and its northern sunshine were foremost in his thoughts. At twenty-one he had more than his share of youthful impetuosity and a streak of adventure.

Dr. Dunmore Lang's books, a present from his Scottish grandparents, had been compulsive reading, fanning the sleeping embers of adventure into flickering flame. The blaze was whipped to its present fervour by the ecstatic accounts of those who, although not much older than himself, had already savoured the romance of the Queensland bush.

The infant colony's first Premier, Robert Herbert, waxed lyrical on the prospects of agriculture and in particular pastoralism, in the new colony. With the Scott brothers and George Elphinstone Dalrymple, a distant connection of Jack's, Herbert was part-owner of a north Queensland property with the alluring name of *Valley of Lagoons.* Pat Leslie, also part of the Dalrymple clan, who with his brother settled on the Darling Downs in southern Queensland, was an antipodean convert and spoke rather disparagingly of the family home when he returned on a visit. Britain was 'too fenced-in. There was no room to turn around in it.' He couldn't swing the proverbial cat and hankered for the open spaces and wide skies of his adopted country.

Jack could sympathise. He felt exactly the same way. Medical School in Edinburgh, though he had an abundance of natural aptitude, he found too hampering, too constricting. He decided to emigrate and here he was, aboard the *Wansfell* with a group of other enthusiastic settlers bound for Port Denison. Now known as Bowen, Port Denison was the furthest port of the new colony of Queensland, well north of Rockhampton which was situated near enough on the Tropic of Capricorn itself. The *Wansfell* was the first of many immigrant ships to dock at the northern port.

Jack had no money to finance himself into a squatter's run stocked with sheep or cattle. His was an assisted passage and his plans were to enter pastoralism as a stockman and to work his way up from there. The wages weren't overly generous, a pound a week and his keep – thirty shillings if he were lucky – but Jack was footloose and fancy-free, his whole life lay ahead and he had no one to consider but himself. He was young, extremely fit, a willing and a good worker and a better than average horseman. What more could any employer want? Jack could already see himself riding behind a mob of a thousand or two prize merinos, boiling his billy at a billabong and eating his meal of cornbeef and damper around the campfire in true colonial style. That'd be the life for him. No more bleak foggy days cramped up in Edinburgh.

The passage up the east coast proceeded at an irritatingly slow rate, following almost exactly the path set by Captain James Cook nearly a hundred years before. From Rockhampton to Port Denison the *Wansfell* seemed to travel even more leisurely despite the prevailing south-easterlies that blew constantly by day. At last the maze of coral islands constituting the Whitsundays was glimpsed but what wasn't so clearly visible was the submerged coral reef the *Wansfell* struck. No matter what strategy Captain Hugh Brodie attempted, the ship remained grounded on the reef. It was a tantalising situation. The mainland was close enough to provoke the captain into sending the passengers in by longboat but the route was so precarious with its uncharted waters and unpredictable currents and undertows that his good sense immediately dismissed the idea. There was nothing to do but wait – and wait they did, while tempers frayed to shreds, food had to be rationed even more conservatively and a high tide came to float them safely off - three weeks after grounding.

The time spent stranded on the reef blunted the elation the young voyagers felt at being so near their destination but time heals swiftly when there's the prospect ahead of a long life of adventure and derring-do. The elusive Whitsundays which, miragelike, tormented them for what felt like an eternity as they sailed north, were slowly drawn abreast of and finally passed, the end of the long sea voyage very near at hand – barring further mishaps.

Jack was joined on deck by other young hopefuls anxiously waiting for the first sight of their journey's end, the culmination of their long dreaming. There was neither pier nor wharf. That came later when Governor Bowen steamed in on a vice-regal visit to honour the town that took his name.

The mangrove lowlands bordering the coastal range gave way to a superb natural harbour almost locked-in by Gloucester Island. It lay to the seawards like, as someone remarked, 'a monstrous saurian'. A sandspit and the backdrop of the coastal range enclosed water that lay silvery calm, protected from the boisterous trade winds that should have sped the ship on its voyage north.

The *Wansfell* hove-to temporarily as some horses were unloaded into the bay and swum ashore, a regular occurrence by all appearances and carried out without fuss or complication. Some men swam in with them. Jack would have liked to have joined them but his shipboard friends weren't tempted. Somewhat disappointedly he remained on deck with them. Small boats came out to ferry the passengers, their luggage and what cargo was aboard to the infant township.

Jack had arrived.

Carrying his baggage he followed the others up the street, past the Lands Office where George Dalrymple began recording land claims from New Year's Day 1861. It was now December 30 1863. The calico tent 'office' had been replaced by a more upmarket hut and Dalrymple, disentangling himself from the tangle of red tape that a Land Commissioner's position brought with it, had moved on. He'd not been popular at the end with his senior, A.C. Gregory, the indefatigable and able explorer who thought Dalrymple should have spent more time in his office keeping the paperwork up-to-date. By 1862 Dalrymple had processed 454 applications for land and granted licences for 144 more over a total of 31,504 square miles of country.

Dalrymple spent his time in the field checking boundaries, settling disputes and seeing that the compulsory stocking condition of the licences was not violated by the simple expedient of driving the same requisite mob of cattle from run to run in order to meet the stocking requirement of each run. Dalrymple was no fool. Although head office in Brisbane found him lacking, the settlers of the Kennedy district did not. Just prior to the *Wansfell* landing they presented him with a silver coffee set – an 'acknowledgement of the many acts of kindness they have experienced at his hands, and as a mark of their utmost confidence and esteem.'[1]

As Jack and his friends went on to their hotel, past the architecturally-similar stores which bought and sold goods as well as acting as banks and agencies, he was caught up in the cheerful bustle and cameraderie of the new outpost. With youthful bouyancy he hoped it wouldn't be too long before he fronted up at the Lands Office to select his own run. Port Denison radiated an aura of optimism and its inhabitants were eager to be part of Queensland's 'colonising enterprise' as Dalrymple expressed it.

The old New South Wales government had earlier tried a settlement at Port Curtis, the site of today's Gladstone, but it expired after a weak and sickly infancy. The Port Denizens, in particular the Kennedy Men who accompanied Dalrymple on his pastoral explorations, were emphatically determined that Port Denison would not follow suit but would, with its magnificent harbour and hinterland, bring credit to them all.

Jack listened quietly that night to the tales of the Kennedy men, the bold bushmen who went exploring the Kennedy district with Dalrymple. He was attentive to those who came after them and the rare few who came before them and who were denied their selections on the grounds that the district was not yet 'open'. The settlement of the Kennedy District was held up for a year by inter-colonial distrust. N.S.W. agreed to open the district for settlement some weeks before the new colony was granted self-government. Tenders for runs were to be accepted after the first of January 1860 but when the Queensland parliament met, some of the Queenslanders became suspicious of what could have been N.S.W. speculators and 'map selectors'. These were pastoralists who took up land merely

1. *Brisbane Courier*, 10 November 1863

by looking at a map rather than by personal inspection. It seemed like picking race winners with closed eyes, a pin and a form guide.

Queensland wasn't born with a silver spoon in its mouth on its separation in 1859. It had only seven pence halfpenny in its treasury at Separation. Land was its only asset and, to cap it all, N.S.W. also presented the emerging colony with a bill for services rendered. The new government aimed to protect the land in its jurisdiction for genuine Queenslanders, individual farmers as opposed to southern companies and speculators. It wasn't until after much argument and another year had expired that Dalrymple was installed in his Lands Department tent at Port Denison. From the records, his mates must have gone to see him in his official capacity straight after (or during) the party he gave to see the New Year in at his Brisbane home. Philip Selheim took up Strathmore, Ernest Henry selected Mt. McConnell and by 9 a.m. January 1 over a thousand square miles of the new colony had been claimed.

Even the Allinghams who came north before Dalrymple's expedition to stake out runs only to find them not yet officially available, returned and made their claims legitimate.

Queensland was the only Australian colony to 'start with full responsible'

Government with two Houses and a Premier as well as a Governor. However the new colony's population of twenty-two thousand was rather politically naïve. It was populated mainly with doers, rather than with entrepreneurs. Queensland's initial area was a handsome 668,497 square miles and this was further increased when her western boundary was extended to take in more of the shores of the Gulf of Carpentaria. Queensland's senior statesmen were right in recognising the worth of the land.

The ambience at Port Denison was of hustle, bustle, continuous activity and a boundless confidence in what the future would bring. On the day that Dalrymple proclaimed Bowen a town, the American Civil War began with the attack on Fort Sumter. The price of cotton needed so desperately by the Lancashire mills increased five-fold when, in America, the southern cotton-producing states were blockaded. Cotton would grow wonderfully well on the coastal plains near Port Denison the agricultural experts decided. So would tobacco, sugar, tea, coffee and almost everything that was currently imported for Australian use. Here was an opportunity to become both self-sufficient and a trader in the exporting world.

Jack's choices doubled. He could rear cattle or sheep on an inland run – assuming he could obtain the capital – or become a planter growing acres of cotton or tobacco on the coast. Apart from the climate being so much better than the bleakness of Edinburgh, the opportunities for a young man with ambition were practically boundless. Hopes were high and idealistic plans sanguinely conceived. There was even talk of gold strikes and fortunes to be made in the Rockhampton area.

THE FRANTIC RIDE OF ROBERT CHRISTISON

Robert Christison, after his first frantic ride down from *Lammermoor* on selecting the property, rode in at a more leisurely pace to arrange for supplies to be taken back there by dray. Several teamsters with horse or bullock wagons were waiting about the Port ready for such contracts. The explorer Landsborough, a fellow Scot from Berwick, assisted Christison in the search for his run and Christison in turn gave good advice to newcomers and offered help to get them established. The tall Scot with honesty and goodwill radiating from his intense blue eyes made an impression on Jack. He decided Christison's counsel would be well-considered and advantageous to an inexperienced hopeful.

Landsborough, an explorer of whom a contemporary remarked that though they'd all done 'a starve or two'[2] none 'starved as cheerfully'[3], had been on an expedition the previous year to search for Burke and Wills. In the course of his journeying he'd come to Towerhill Creek, one of the far-northern headwaters of a stream entering Lake Eyre. There he'd found a large area of good grazing land and waterholes big enough to float a steamship. He'd marked the place with a tree blaze and an arrow together with the date he'd camped there, March 22nd 1862.

Following Landsborough's instructions Christison found the area his friend described and set about making a small horseyard while his hobbled-out horses recovered from their trip. All the while he blissfully thought he was far enough from civilisation and other landseekers not to have to hurry away posthaste to record his selection. He was wrong.

While cutting posts and rails for the yard a stranger rode up, asked 'a great many questions' and stopped mid-sentence as if a thought had suddenly come to his mind.

"Are you working for Christison?" he asked.

"Yes." That was no lie."Do you want to see him?"

The stranger inexplicably laughed and left, promising to see Christison later.

Suspicions roused, Christison summoned his horseboy to get the horses in, thankful that they'd had a short spell, brief though it was. Quietly they set off and rode to the main range. From there, Christison got a daylight start leaving his

2. A.C. Gregory
3. A.C. Gregory

offsider, Gailbury (Dalleburra), to follow at his leisure while he set off on his stouthearted, part-Arabian stallion Jack Straw. The stallion came to Christison with the reputation of an arrant and incorrigible buckjumper. There was instant rapport between them and Jack Straw became a paragon of virtue and endurance. In a momentous feat for horse and rider they covered the distance from Lammermoor to the Lands Office, considerably in excess of 300 rough miles in four days and applied for the occupational licence that would legally secure the land until Christison was able to stock it and apply for a lease. The requirements were for five head of cattle or one hundred sheep per square mile. He intended to stock with sheep. There was no market in the north at that time for beef. Unlike meat, wool was not perishable and could be stored and freighted out by wagon with relative ease.

After several days of hard riding the stranger arrived at Port Denison and fronted up at the Lands Office but as he hadn't been spurred on by Christison's determination, he lost the race for the station. It was already taken up and the loser was left puzzling how one man on one horse could have accomplished such a Herculean feat. Proving such marathon rides weren't impossible, in 1867 another equally determined man followed suite when J.G. Mac Donald rode from the Gulf region to Port Denison in twenty-eight days, a ride of just under 1600 long miles.

Jack decided he'd try to get an introduction to Christison. Scotsmen stuck together. They'd have common friends and acquaintances. Maybe he'd offer employment and the opportunity to gain valuable pastoral experience. Enthralled by the romance of it all Jack listened to the talk of the runs and the distances involved, both, even with his previous experience of Australia, almost inconceivable when compared to English or Scottish conditions. He was extremely interested when told of a scheme proposed by the Scotts of the *Valley of Lagoons* to provide well-connected young men from 'Home' practical working experience with sheep and colonial life.

Jack was also attracted by the name of the station. He saw in his imagination the endless miles of grazing land interspersed with broad lagoons teeming with waterlilies and waterbirds, and the suggestion that at the end of a reasonably short apprenticeship the new chums could obtain both land of their own from the *Valley* plus sheep to stock it. He also knew of the Scotts by repute.

With his mind awhirl with these intoxicating plans he went to consult one of the storekeeper agents – and had his dreams cut short. Like almost everyone in the colony, the Scotts were in financial difficulties and the colonial experience gambit was introduced in the hope of raising hard cash. The gentlemen-stockmen were to pay for their education – in advance – and be recompensed at the end of their training with pitifully small parcels of land and the token sheep. Jack's financial status –barely enough to keep him afloat until he got a job – ruled him out. Disappointment bit hard as his dreams collapsed.

DR JACK WALKS TO THE GOLDFIELDS IN CALLIOPE

The goldrushes proved a boon to both Jack and the Scotts. The latter were left with their land and their flocks undiminished as the new-chum jackaroos unceremoniously deserted, succumbing to the allure of instant riches on the goldfields and leaving the money for their pastoral training in the Scotts' bank account. For Jack and his new-found friends the conversation turned to gold. Though they remembered the talk of the Canoona rush inland from Rockhampton prior to Separation when only the fortunate first-comers made any money and the late-comers received nothing but bitter experience. The Calliope field was working-out well and looked like being a long-term proposition. At least miners were making wages greater than that offered to pastoral employees. Earlier in the year gold was found to the north of Rockhampton. There appeared to be gold everywhere in that country waiting, even anxious, to be found by the first audacious prospectors who ventured out into the hills and gullies. It was now 1864, the twenty-eighth year of Her Majesty's reign. Abraham Lincoln was President of the United States and General Sherman had entered Atlanta.Maximilian accepted the throne of Mexico and in Australia Ben Hall was holding-up coaches and squatters. Herbert was Premier of Queensland where the canegrowers had just processed their first commercial sugar and Lantern won his Melbourne Cup.

The time was right for making decisions. Some of Jack's shipmates elected to try their luck at Calliope. After a nightlong discussion of hopes and dreams, hearing his friends put forward the pros and cons of various choices applicable to the free spirits that they were, Jack settled for life as a goldminer. When daylight came he rolled his few belongings into a passable swag and all thoughts of Robert Christison and a career in pastoralism vanished. When the would-be gold diggers left the camp at daybreak the next morning, singing boisterous gold-mining songs to boost their spirits, Jack was with them.

A horse would have cost him twenty to thirty pounds, more than he possessed and the equivalent of almost three months wages on a station. Jack walked. He had plenty of mates and trudging along together, chatting, exchanging experiences and occasionally singing a popular marching song to spur them on, they could always, when all else failed, pull up near water and put on the billy for a tea reviver.

Jack stowed his possessions rolled neatly in his swag, safely embedded in the centre of which was a smaller roll containing his rudimentary medical kit. His mother, deeply disappointed that her tall, handsome son wasn't to carry on as a doctor in the Old Country, insisted he take his tools of trade with him.

"They'll come in handy," she contended."It'd be a shame to waste those years at medical school."

Like most mothers, at least some of the time, she was right but Jack pined for a more adventurous life than doctoring. In the roll beside the medical kit, swathed in a piece of silk his mother had also given him, was his favourite pistol. As the group moved south they sometimes met others working north, disappointed with claims that poorly rewarded their hard work and high aspirations. At first word of a new strike – even the likelihood of one – luckless prospectors were instantly off to try their chances again.

The heat, especially to someone like Jack, fresh from a bleak Scottish mid-winter, was over-powering. Trudging along the narrow track that wound in and out of tall trees and underbrush, up and down according to the inclination of the land, there was little breeze to fan the travellers and the summer heat beat down upon them with the force of a brass-beater's hammer. Smoko breaks came more often even if the strength of the tea had to suffer as a result and boots were eased off blistered and aching feet at every stop.

It was also the time for the severe electrical storms that herald the coming of the North's monsoon season. One of the men had a small canvas tarpaulin that would partly shelter four mates if they jammed together like sardines in a tin, slung the tarp barely high enough to allow them to sit up and took the precaution of digging a moat around their castle to drain off storm rain before they unrolled their swags. Bill, the owner of the tarp, had been chasing gold for some time and was equipped, if not with 'wheels', with a wheel in the form of the ubiquitous wheelbarrow. In appreciation of a dry-ish camp at night, Bill found sufficient voluntary labour and complimentary pipefuls of tobacco to travel in comparative luxury with someone else pushing his barrow. The barrow also carried his pick, shovel and gold dish, items Jack had yet to procure.

Calliope was their destination but should they meet with another rush en route, their minds were open to persuasion. There was a path of sorts but mostly they simply kept the rising sun to their left and headed south. They passed the track that led to the Peak Downs strike of 1862 dismissing it as a minor field, not knowing – and probably not caring if they had the knowledge – of the great coal field it would become. Nearing Rockhampton they passed more huge coal reserves at Blackwater, the gem fields of Anakie, Sapphire and Rubyvale and the undiscovered goldfields of Yaamba and Ridgelands.

They scorned the turn-off to the much-maligned Canoona field only thirty-five miles upriver from Rockhampton and unwittingly passed within a few miles of that mountain of gold, Mount Morgan. Canoona fired the hopes of tens of thousands of Victorian diggers eager for instant riches and success and filled the pockets of the boat owners who shipped them there – and back. Although over 5000 ounces of gold were taken out in the first few months by the early comers, thousand of the late-comers straggled back to the infant Rockhampton considerably sadder and wiser but not the slightest bit richer.

Jack's companions were happy to leave Canoona to its ghosts but a couple, feeling lucky, split from the party to seek their fortunes around Cawarral where talk of 'colours' being found was reported on the grapevine. The remaining friends didn't dwell in Rockhampton. Their pockets were all but empty. So were their tucker bags and after securing more tea, sugar, rice and flour they took the southern route again heading for Calliope. On the road Jack's skill with the pistol was the subject of practical admiration. He rarely wasted a shot and pigeon, duck and wallaby were frequent additions to their otherwise very basic menu. Jack soon learned another skill and found he had an aptitude for it – camp cooking. His dampers and the smaller johnnie cakes cooked on hot coals were as popular as his stews.

DR JACK AS MEDICO OF THE CALLIOPE GOLDFIELDS

He didn't know what to expect when they reached their destination. In Scotland, even the smallest fishing village had a look of solid permanence evidenced in sturdy stone walls and squat roofs. Port Denison, gateway to places with intoxicating names like the *Plains of Promise* and the *Valley of Lagoons*, with its roughly made timber structures, looked vulnerable and temporary to the extreme. With the discovery of superior building clay nearby, one or two of the new buildings were of the more enduring brick. In the early sixties, Rockhampton, built on the broad reach of the Fitzroy, though a little older was only slightly more substantial. Calliope, when they reached it, was of the ephemeral category. Canvas and bark appeared to be the architect's choice of the month. There were a few more solid buildings – a bank, a store and the ever-present pub.

Only Jack and Bill were left of the original group. Their partnership was augmented with the arrival of a veteran from the Victorian diggings who'd joined them in Rockhampton. He was sick of trying his luck in lonely gullies without a mate to watch his back in case of a stray but accurate spear. Paddy had spent two days and a night in a cave-in before he was able to scrape and scratch his way out and felt that, for him, prospecting without a mate had lost its charm. Jack was appalled at Paddy's suppurating leg wound, a great gash to the shin bone caused by an unfriendly lump of quartz in the cave-in. Despite Paddy's protests that it was on the mend, Jack got out his tools of trade and cleansed the wound as best he could.

"It should've been stitched," he commented, but, realising that the time to initiate that procedure was long gone, he patiently burnt back the long weal of proud flesh with blue stone This was done a little at a time, night after night, by the light of whatever campfire served their purpose. Finally a long jagged scar like a red snake was all that remained of the wound on Paddy's milk-white shin.

Paddy swore Jack saved his life, a rather exaggerated claim, but touching. Jack, usually a modest man, couldn't resist a slightly smug smile. The Irishman took a fancy to the young Scot – or native Australian if Jack's gaelic-accented claim was true – and a Protestant at that.

Jack was a man of few words, reticent, yet he was at all times aware of what was happening around him. He took a sincere interest in the men within his circle and was always ready to offer a helping hand. Paddy appreciated Jack's kindness of manner and his skill. He also admired Jack's lithe athletic body and the physical endurance he displayed each long day of their journey but his affection didn't blind him to the fact that's Jack's knowledge of gold prospecting and mining could be writ large in tar on a toe-nail. Mining was a science Paddy knew something about. He'd teach the young fellow.

Paddy was also interested in Jack's choice of the copper sulphate or bluestone to use on the proud flesh.

"An old soldier's trick," Paddy announced around the fire one night.

"Yes," Jack agreed. "It's often used on neglected gunshot wounds." He left it at that while Paddy and Bill discussed mines and horrific mining injuries until the billy was boiled for a final cup.

Staring into the flames like a gypsy engrossed in her crystal ball, Jack began talking in a soft monotone of the misfortunes of the nationalist Juarez in Mexico. A Zapotec Indian, Juarez had, as an elected President, tried to install liberal reform in Mexico and resisted the French occupation of his land a few months before Jack set sail for Port Denison. The French, Spanish and English set up an Imperial parliament of people opposed to Juarez and his democratic ideals and offered Maximilian, Archduke of Austria and brother to the Emperor Francis Joseph, the Mexican throne. In the meantime, Juarez and his supporters were driven to the northern border with orders – said to be from Maximilian – that if captured they were to be treated as bandits. In other words, they were to be shot on sight.

Jack was tangibly shocked at the possible fate of the patriot. He liked to see justice done. It was no crime to defend democracy in the land of one's birth. France, England and Spain chose to support the interests of their bondholders over those of the Mexican people. Only the government of the United States of America refused to recognise the new emperor.

So ardently did Jack defend Jaurez and his La Reforma government that his two companions wondered how their dour young Celt could be so quickly transformed into an impassioned firebrand. They were soon to learn that Jack didn't do things by halves. Anything he believed in he espoused with a romantic fervour greatly at odds with his usual withdrawn manner. One reason, he confessed, for deserting pastoralism was to make enough from gold to enable him to go to Mexico and to help restore Juarez and his vision of democracy for the Mexican people. The following morning Jack was first up to stir the slumbering campfire to life and to get water for the tea billy. Benito Juarez and the refugee Juarists seemed to be forgotten.

Once the party set up camp on the goldfields Jack's medical skills quickly became common knowledge. With Paddy's tuition, his mining competence brought him almost enough to pay his share of the modest tucker bill but his nights were spent treating the sick and injured. He could not bear to see a fellow man suffer and not go to his aid. Broken limbs were splinted, open gashes caused by fights were stitched, neglected wounds disinfected and dressed, quinine dispensed to the fever-stricken, laudanum given to those suffering the hells of dysentery and a dose of sarsaparilla or lime juice given the victims of scurvy. Sarsaparilla was the specific for scurvy, a severe dietary deficiency disease and the price of a dose that would give instant relief was five and a half pennyweights of gold.

What patients were financial enough to pay their doctor did so in gold, carefully measured on the little portable scales all seasoned miners carried in their gear. Those who were unable to pay were treated for nothing and sang 'Dr.' Jack's praises loudly for the rest of their days.

Miners rarely became fanatically attached to any one field. A slackening in the amount of gold found or the slightest whisper of a strike somewhere else was enough for them to take to the road again forever seeking that pot of gold at the rainbow's end. Jack prospected the creeks and gullies, at first under Paddy's tutelage and later, as he gained competence, on his own, until he followed all the rushes in the Rockhampton hinterland, visiting the town and its quickly growing society on rare occasions.

DALRYMPLE AND JARDINE SLUG IT OUT IN ROCKHAMPTON

While Jack was slaving away on the goldfields Dalrymple often had occasion to visit Rockhampton. No longer the end of the world, Rockhampton now played host to travelling minstrel troupes, acrobatic performers and even a circus. Crinolines were in fashion and hoop-skirted ladies bedecked in ostrich feather plumes, velvet jackets and parasols were a common sight. Unfortunately, George Dalrymple, a man who loved the good things of life as well as the vocation of an austere explorer, was seen once too often escorting a particular lady to balls and socials in the absence of her husband. Rumours circulated wildly, some originating from a policemen who Dalrymple thought was spying on him. He complained to John Jardine, the local Police Magistrate and father of the two young men Alec and Frank Jardine who had successfully and despite great difficulties overlanded to Somerset at the tip of Cape York Peninsula. Politely he asked that Jardine investigate and stop these 'slanderous charges'. The lady's husband was also justifiably annoyed by the rumour-mongers and, provoked by a local Justice of the Peace, gave that gent a good old-fashioned horse-whipping.

The scene was set. As Jardine was riding home one evening, Dalrymple came up beside him again demanding that he do something about the slander. Name-calling took over from good sense and finally the enraged Dalrymple took a leaf from the husband's book and struck Jardine over the head not with the carved ivory handle of the whip he carried but with the lashed end. His sense of timing might have been justified but his sense of place was definitely wrong. The fracas occurred barely thirty yards from the local police station. Jardine called out the gendarmes and Dalrymple was promptly locked-up for the night. The press had a great time of it. All four contestants were Justices of the Peace.

At the first hearing an Irish witness used such picturesque language in describing Dalrymple's assault on the Police Magistrate that the whole court was in uncontrollable tears of laughter. It took quite some effort to call the session to order.

Neither Dalrymple nor the aggrieved husband would break their silence, for to do so would bring the lady into it all and to protect the name of the woman concerned was their dearest desire. The husband was fined a fiver and Dalrymple remanded until the next session, in April 1864. Many in Brisbane, unappreciative of Dalrymple's informal ways and hoping the pesky Northerner would get his come-uppance, predicted his fall from grace and a lengthy jail sentence.

Dalrymple's friends were concerned too. The people of Rockhampton and the far north were fully behind their hero. Jack agreed with them. Dalrymple could do no wrong. He was merely upholding the honour of the weaker sex.

Dalrymple himself, with work to do and unable to stand the waiting, left for the Valley of Lagoons from where he, nine white men and two Aborigines set off to find a suitable dray road down the Seaview Range to the port of Cardwell on Rockingham Bay. A marathon task. To show a confidence they didn't really feel they made the attempt with three bullock drays, sixty-odd working bullocks, eighteen horses and a small mob of sixty-three fat cattle destined to feed the settlers hungering for fresh meat supplies.

Their progress was heartbreakingly slow. This was the kind of country that broke Kennedy's heart almost as soon as he started his tragic journey. Dalrymple and his men had longer experience with northern conditions and were considerably more practical. On April 24th the Cardwell settlers heard whipcracks, looked disbelievingly towards the range and there was Mr. Dalrymple and his party safely arrived. The *Valley* now had a usuable port and the port a valuable hinterland to support it.

Dalrymple got there just in time too, for while he'd been held up on the perilous climb down the Seaview range, his case was heard with only his lawyer appearing in his defence at Rockhampton. The verdict was for Jardine. For his hot-headedness Dalrymple was forced to pay a colossal five hundred pounds damages bill.

It didn't affect his popularity. The townspeople of Port Denison had made their presentation and didn't regret one word of it; the policeman in question was transferred but Dalrymple's friends had plenty of opportunity to agonise over the outcome while the man himself was battling his way through the inhospitable ranges to the coast.

Jack was pleased to see that Queensland, more especially the North, appreciated its heroes. Heroes were something Queensland had in numbers. The goldfields were good training fields in endurance, optimism and mateship and Jack couldn't see why a man should be punished for defending a woman's honour. Years later, rumours circulated about Jack concerning a duel fought in Rockhampton. Perhaps he did fight a duel for he was an excellent swordsman and held his belief in honour high but maybe the story got mixed the way the best romantic tales do.

Over the years Jack spent in the Calliope-Rockhampton district he met men who would remain his friends for life. At Ridgelands, to the north of Rockhampton, then on his own after serving his apprenticeship with his Irish friend Paddy, Jack panned for gold along with another solitary miner, Jack the Hatter, John Murtagh Macrossan. Like Jack, Macrossan was a man of few words but with passionate beliefs. Born in Ireland, Macrossan was sent to Glasgow at the age of sixteen to further his education. He spent several years there and acquired a Scottish accent along with his education. Macrossan became a lifelong friend and a parliamentary ally.

Ridgelands was also the setting for a saga into which Jack was later unwittingly drawn. Halligan, a well-known gold buyer from Ridgelands and Morinish, was set upon by two men, Palmer and Williams, who demanded his valuables. Halligan slashed Palmer across the face with his whip and galloped off but the two desperadoes gave chase and Palmer brought Halligan off his horse with a shot to the chest. They tied and gagged him, took the notes and gold and a gold ring from his finger. They callously dumped him in the dense bushland out from Rockhampton and abandoned him there to die.

Their accomplice in Rockhampton, a hotel-keeper named Archibald, shared out the booty, chopping the retorted gold into three equal pieces. They tossed for Halligan's signet ring. The next day, on going back to check on Halligan the trio found him dead, put his body in a bag they'd brought for the purpose, weighted it and threw it into the river.

On the way back to his young wife in Gympie, Palmer gave a barmaid an ounce of gold he'd chopped with a tomahawk from his twenty-four ounce share of the gold. She showed it to a policeman acquaintance who, knowing Halligan had been carrying retorted gold, became suspicious. A two hundred pounds reward was offered and Rockhampton citizens quickly subscribed more than double that amount to add to the official sum. Archibald, the accomplice who dragged the body into deeper water so that it would sink, informed on Palmer and Williams.The murderers were on the run.

RACE RIOTS AT CROCODILE CREEK AND THE FABULOUS MOUNTAIN OF GOLD AT MT. MORGAN

Jack followed the gold trail to Crocodile Creek. Here Chinese miners flocked in their thousands.The men already on the field were so irate at the intrusion of so many Asians that in 1865 it became the first Queensland goldfield summarily to ban Chinese from the mines by an act of Parliament after riots and disturbances broke out between the two races. Crocodile Creek wound around the foothills of Mt. Morgan and fossickers again panned for gold within a few miles of that legendary mountain without suspecting what riches lay beneath its surface.

Gold in Australia was traditionally associated with quartz until, when Charters Towers burgeoned into the self-styled 'World' of its miners at the beginning of the 1870s, gold was found in ironstone. The initial find however was made in a quartz outcrop by an eleven year old indigenous horseboy, Jupiter Mosman who made the discovery while bringing the horses back to camp. The gold at Mt. Morgan was encased in ironstone and orthodox prospectors shunned the area until Charters Towers re-wrote the geological rule book.

In the sixties, Mt. Morgan was a cattle run and stockmen kept a vigilant watch on the cattle in case hungry diggers fancied fresh protein in their all too bland diet. One stockman, Will Mackinlay, camped out in a bark hut, had time to do a little fossicking while still keeping an eye on the herd. He found copper, a little silver and gold in the mountain ironstone but was thwarted in his continuing search by a settler John Gordon and his sons taking up a square mile of the valley and the mountain where Mackinlay had been prospecting.

The Gordons fenced their run but never noticed the glint of gold as they sunk postholes in the ironstone. Though sitting on one of the richest parts of the continent they didn't prosper and after six years moved on. By 1882 the mountain was again deserted. One of the Gordons, married to a Miss Mackinlay, worked at Galawa goldmine for the Morgan brothers. He was something of an alcoholic and the Morgans, tiring of his lack of responsibility, sacked him. Desperate, Mrs. Gordon promised that her husband would not only mend his ways but would show his bosses a silver prospect if they would only give him his job back. Sympathetically the Morgans re-instated Sandy Gordon and sober, he took them to the 'silver lode'.[4]

The Morgans, experienced miners, crushed some ironstone on a shovel and washed the samples. Their expectations were low but to their astonishment, there was more gold than dirt in the dish. However the gold was very fine and for all

4. Cyril Grabs, *Gold, Black Gold and Intrigue,* CQU Press, Rockhampton, 2000

their experience they still weren't fully convinced. Where was the quartz? There was just ironstone and no lode could be that rich! There was too much of the yellow stuff to be the genuine article. They had enough faith to buy the land, abandoned by the Gordons who considered it poor cattle country, for a pound an acre. Then they had their samples re-tested. It was gold. Some say the early crushings, not terribly efficient because of the unknown qualities of the ironstone, yielded an average of ten ounces to the ton. A phenomenal amount.

In 1889 a ton of gold a month was being recovered and pound shares in the Mount Morgan Gold Mining Company jumped to eighteen pounds a share – if any of them were available for sale. Fortunately for science in Australia, one of the principals of the early company was T.S.Hall. He brought in his brother Walter and used profits from the Mount Morgan Mine to found the Walter and Eliza Hall Trust which has fostered many worthy projects and graduates over the years. Some seventy years on Sir Frank Macfarlane Burnet did work on poliomyelitis viruses and influenza viruses and vaccines while director of the Walter and Eliza Hall Institute of Medical Research in Canberra. For his research into 'immunological tolerance' he was joint winner (with Sir Peter Medawar of the U.K.) of the 1960 Nobel Prize.

But by the time the eighties arrived, Jack and his mates had deserted the Rockhampton hinterland for the richer prospects at Gympie which were left, in turn, for newer fields in the Far North and for Jack's entry into politics.

2

GOLDEN GULLIES AT GYMPIE 1876

Collapse of Agra Bank. New colony in dire straits until gold is discovered at Gympie. Jack follows miners to Gympie.Acts as doctor.Meets Mulligan, Philp and Dean Matthews. Becomes local hero, member of committee to draw up local mining laws and elected to the Miners' Court. Father Horan, drunken miners, the Aborigines and Johnnie Campbell.

SEVEN PENCE HALFPENNY IN THE QUEENSLAND TREASURY

When Queensland became a separate colony on December 10th 1859 with its massive sum of seven pence half-penny in the Treasury, it also had an entry in the Statistical Register for December of that year compiled by order of Governor Bowen which stated under the heading Public Debt: 'Not yet adjusted with New South Wales.' The Mother Colony wasn't going to let its newest infant start off with a clean slate – as well as its seven pence half-penny. Loans had to be undertaken to secure finance to run the new colony and it was hoped, with the advent of government charges, they would soon 'trade' out of their financial trouble.

The building of railways, rather than their closing-down which is more popular today, was expensive and even the pier to be erected at the township of Bowen in readiness for the Governor's visit was to cost ten thousand pounds that the colony didn't have. Income was in very short supply and the budget wouldn't balance no matter what was tried. One Premier, Macalister and his treasurer J.P.Bell had the idea of a 'greenback' currency, unsecured notes printed by the Government but with no gold reserve backing, as an expedient measure to keep public servants paid and the trains running. Governor Bowen disagreed. He had the final say.

Money was eventually borrowed overseas. A temporary reprieve only as the ink on the mortgage was barely dry when the main lender, the Agra and Masterman's Bank, failed. Queensland's fifty thousand pounds a month that was to come from that bank's vaults didn't eventuate. Then the Bank of Queensland, another infant, died suddenly if not unexpectedly. Anyone unfortunate enough to be left with Bank of Queensland pound notes considered themselves very lucky indeed to be able to get rid of them at half price. Work on the highly-regarded railway to Ipswich and Toowoomba was cancelled. Unemployed railwaymen walked the Brisbane streets unsuccessfully looking for work. Queensland was worse than broke. It was insolvent.

In desperation Treasury Bonds were issued, redeemable in 1869 at a very generous 10% interest and for, in total, two hundred thousand pounds. Investors were slow to respond. The Government slumped back exhausted and hoped to goodness something would happen to save them. A miracle would be most welcome.

The Canoona rush, though pre-Separation, was a failure and the Rockhampton gold finds did little for the colony's coffers but to bring in more mouths to be fed. Louis Hope received a land grant in recognition of the experimental work he'd done growing cane and producing sugar in Queensland but the colony needed something much more spectacular to solve its problems. It needed a Five Star Gold Rush.

It advertised a reward and almost by return mail got news of a sensational gold strike. Late in 1867 a prospector, James Nash, went into Flavelle Bros. Jewellers in Queen Street to sell seventy-five ounces of gold. Co-incidentally, in the shop at the same time was The Hon. W.H. Walsh, Member for Maryborough. Walsh asked the miner where he'd found the gold. Rather cagily the man replied, "Up north". His reply led to the rumour that the gold came from Northern Queensland, probably from the new Cape River field.

The gold was put on display in the jeweller's window and the rest, as they say, is history. The gold came from what was to become the Gympie goldfield.

Sir George Bowen promptly proclaimed the area.

> *The Upper Mary River Goldfield, 40 miles due South from Maryborough, 25 square miles, commencing at a point bearing South and distant one mile from the junction of Gympy (sic) Creek with the Mary River; and bounded thence on the South by a line bearing West*

two miles, on the West by a line bearing North five miles, on the North by a line bearing East five miles, on the East by a line bearing South five miles and again on the South by a line bearing West to the point of commencement.

Given under my hand, at the Seal of the Colony, at Government House, Brisbane, this thirtieth day of October in the year of our Lord one thousand, eight hundred and sixty-seven in the thirty-first year of Her Majesty's reign,

G.F.Bowen

By His Excellency's Command.

E.W.Lamb.

GOD SAVE THE QUEEN

God had also saved Queensland – at least for the time being.

Naturally, word travelled fast and even before the field was officially proclaimed the miners flocked in from other fields where their hopes were fading. Diggers came from the townships further north; with work on the railways curtailed, ex-navvies deserted the streets and footpaths of Brisbane; men arrived from the open plains of the Darling Downs, from Victoria, New South Wales and even from the Land of the Long White Cloud, New Zealand.

NASH AND THE GOLD RUSH TO GYMPIE

Spontaneously Jack joined his mates in the rush to an instant gold fortune. This time he didn't walk, he rode. He had an eye for a good horse and was taken with a big, bay thoroughbred a young squatter had sent up to Rockhampton. The squatter was no horseman and the thoroughbred soon found that out. Each time the man mounted him, with a few well-placed bucks, the horse would throw him.

Feeling sorry for the young fellow, Jack sprang into the saddle to take the sting out of the animal but instead it behaved like an angel – until the squatter re-mounted. Picking himself up yet again from the dust he asked Jack to take the accursed animal from his sight. He didn't want to see it again. Jack swiftly obliged and he and his new mount were speedily out of the squatter's vision.

James Nash didn't have to go far to find his gold. He'd been at Nanango about eighty miles to the south-west, at the head of the Brisbane Valley. Some time before he'd overlanded from Calliope where luck hadn't been with him. Though in one way it had. Like Paddy, he was buried in about six feet of earth when his workings fell in on him. Fortunately, other miners were at hand and soon rallied to his rescue and dug him out. He was on the point of suffocation but otherwise unharmed, no fractures, no serious bruising.

Zac Skyring, on a trip to buy sheep at Nanango for his Brisbane butcher shops, met in the course of his business with the shepherd caring for the sheep, an old Victorian digger. [5] He talked so much of the gold he'd found that he temporarily converted Skyring to prospecting. They found gold, took it to Flavelle's where, because of the assay, they received a premium price for it and took up a claim surrounding their find. The heaped dish of gold was displayed in the jeweller's window with a card stating 'From Nanango Goldfield.'

Nash was one of the first to this field but the auriferous patch soon petered out and Nash quietly packed his gear and left the field alone, as was his custom, heading north towards the port of Gladstone. He didn't get to his destination. En route, between Widgee and Traveston sheep stations he decided to try a gully that attracted his keen prospector's eye. He got more than enough gold to whet his appetite but in doing so broke his miner's pick. There was nothing for it but to go to Maryborough and purchase another.

No one there was interested in his gold. There had been too many false hopes raised. Nash said times were so bad they'd even forgotten what gold looked like but finally he was able to persuade a sympathetic store-keeper to exchange his gold for tools, tucker and a pound in cash. Refitted, he returned expectantly to his prospect where, back at his creek he picked up a few small nuggets while washing the alluvial and in six days had the seventy-five ounces that inspired the big rush.

As far as the out-of-work railwaymen were concerned, the 'up north' sign in Flavelle's window meant Nash's strike at Gympie and they hurried there as best they could.Some had been employed on the Government's road programs for which they were paid only two shillings and six pence a day. They had nothing to lose. Thousands flocked to the field. Surprisingly many women went too. Not only the flashy Palmer Kate bar-girls of a later rush but also the respectable wives of the miners. They saw it as a God-given chance to make something of their lives. They packed their belongings, dressed the children for the road, the more financial steaming by boat to Maryborough and overlanding from there but the majority went the slow way by dray or even on foot up the Valley road. A 'good' trip took two weeks in a two-horse dray.

The long and multiple skirts of the day hampered them in the outdoor lifestyle. Many a long skirt caught fire when wafted suddenly into a campfire. Wool was a popular material particularly for the less well-off. Long-lasting and relatively

5. *A Retrospective Reverie, Gympie Times,* 16 October 1917

cheap, it had the benefit of smouldering rather than of bursting instantly aflame and probably saved quite a few women settlers from mutilation and death, a suttee before their time.

Crinolines were still in fashion and Catherine Nash,[6] wife of Gympie's discoverer, commented on the ultra-fashionable who added extra width to their hoops and 'looked like beautifully decorated barrels wobbling along on springs.'

Nashville, one of Gympie's early names, had all the hurly-burly and noise of a fairground. The miner's first priority was to stake a claim. On it his future depended. Housing and home comforts were low on the list. Tents and bark humpies were shelter enough until fortunes could be made.

Though meat was in good supply and for sale at a reasonable price, other items were non-existent or in short supply, wrapping paper for one. It was common to see citizens taking home their week-end roast, holding it by the long wooden skewer the butcher drove through it and keeping the flies off as best they could. [7] Coffins also were hard to get and were usually rough affairs, lined with calico and covered with a black crepe drape preserved carefully for the next sad occasion.[8] Diptheria was rife and though Dr. Jack and his fellow medicos did all they could with the limited drugs available, many of the coffins were pitifully small in size.[9]

There were rat and mice plagues which helped spread diseases more rapidly. Cats, when available, were at a premium, a good mouser fetching ten shillings or more – the equivalent of forty pounds weight of choice Wide Bay beef. Snakes were also a problem especially as everyone lived as closely as possible to the river and waterholes.

Queensland, though praying for this miraculous gold strike, had done nothing to frame its own mining laws. It had, in its first year, repealed the Gold Export Duty Act but its mines were still governed by the New South Wales Goldfields Act of 1856. This Act tended to encourage financial investors rather than provide incentive for the individual miner and the Queensland diggers agitated against this. By the time Nash found his Gympie gold, the Gold Export Duty Act was re-introduced (in 1864). It put a fee of one shilling and six pence on every ounce of gold which was worth, in round figures, about three pounds. This, too, galled the miners. The heady days of Eureka were still with them.

---

6. Catherine Nash, *Gympie Times,* 16 October 1917
7. Robert J. Pollock, *Overland with Dray and Team, Gympie in its Cradle Day,* Gympie & District Historical Society, 1917 and 1985
8. Edward Alexander Poulton, *Gympie in its Cradle Days,* Gympie & District Historical Society, 1917 and 1985
9. Although Jack was not a fully qualified medical doctor, his paramedical practice on the goldfields of Queensland was widely recognised and he was referred to as 'Dr. Jack'

A quaint and colourful rule of the 1856 Act was that, on finding payable gold, a digger was required to fly a red flag, no less than a foot square, above or adjacent to his claim. Failure to comply could lead to forfeiture of the claim. In its heyday Gympie was absolutely aflutter with thousands of these red, butterflylike flags.

BUSY DR. JACK IN GYMPIE: MEDICINE AND THE MINING RULES

Not long after Dr. Jack came to the Gympie field and his medical skills became known, an unidentifiable man of about thirty-five was found comatose, possibly dead. The Catholic priest was found and a search made for a doctor. That was something of a problem until some of the miners from the Central fields suggested Jack. He'd doctored them rather successfully at Ridgelands and Calliope.

He was quickly summoned from his new workings and pronounced life extinct. The man was buried and a small post-and-rail enclosure made around the unfortunate man's grave was designated as the cemetery. The local newspaper recorded in amazement the odds of coming upon a doctor so early in the life of the field – 'a real doctor, a doctor in moleskins'.[10]

The following week a man was found dead at the door of Greathead's Spirit Store. The same moleskinned doctor was called and found the man died of natural causes, a 'result of constitutional weakness'. This man was identified by his mate who offered the opinion that had his deceased friend been able to summon the strength enough to enter the store he 'would have survived a kick from a donkey'.[11]

By this time Jack, now a young man in his mid-twenties, had gained a following. He still worried about Benito Jaurez battling it out in Mexico, for though the Juarists had defeated Maximilian's forces at Queretaro in May and Maximilian was shot with his generals Miramon and Mejia after a plea for clemency failed, life wasn't all plain sailing for La Reforma and President Juarez. Jack's friends, however, had other plans for him.

Jack had been a lively child in Victoria when tales of the valour of the miners at the Eureka Stockade were part of everyday conversation and to young boys especially, the miners and their leaders were heroes of the first order. Jack was a man's man, fiercely honest and independent, someone miners felt they could trust, especially as along the way quite a bit of Paddy's knowledge of the yellow metal had rubbed off on him. When the miners called for a local committee to

10. *Gympie Times,* 16 October 1917
11. *Reminiscences by 'A Participant', Gympie Times,* 16 October 1917

draft a set of rules for the Gympie field, more like the Victorian guidelines than the unpopular N.S.W. Act where the Gold Commissioner rather than the miners made the rules, Jack Hamilton's name was put forward. In 1868, with Frederick Lord, James McGhie, James Fisher, E.G. Milligan, Frederick Goodchap, Thomas F. Browne and Hugh Goodwin under the chairmanship of H.E. King, Gold Commissioner, later M.L.A. and Speaker of the House, Jack helped draft regulations for the control of mining in Gympie.

Commissioner King was an imposing figure of immense dignity and impressed the rough-and-tumble miners with his long, silken frockcoat and tall, black, bell-topper hat. The diggers were more soberly attired in a uniform of moleskins, crimean shirt and while working, a fly-veil. The flashy ones, doffed the fly-veil and put a silken cummerbund around their waists for evening wear.

The Mining Rules were authorised by the Minister for Works, Arthur Hodgson, in early October of that year and were so relevant to the field's needs that they remained in force until 1874 when a Queensland Goldfields Act was passed. The Rules of the Gympie Local Mining Court took a bit of drafting and were comparatively as long as their name – sixty-eight pages. Bernays, writing about them in his book on Queensland politics says, 'a feature that strikes one on reading these rules is that the grammar in use in the year 1868 was materially different from that in vogue in 1919, there being several instances in which the nominatives and verbs are in strong disagreement with one another.' [12] But – it did last and served the purpose well for six exciting, action-packed years and, with the grammar, maybe they were just a wee bit ahead of their time.

The Local Mining Court was Dr. Jack's entrée to public life. He still found time to work his own mine but his 'spare' time must've been exceedingly crowded with tending the sick and with politics.

THE HANGINGS OF GOLD COMMISSIONER GRIFFIN AND PALMER IN ROCKHAMPTON

Jack and his mates left the Rockhampton area at the right time for in June 1868 the Gold Commissioner, John Thomas Griffin was hanged for murdering two troopers, Power and Cahill, on the Mackenzie River and taking the money paid for the gold they'd been escorting. Griffith's main downfall was a passion for gambling and a liking for the high life. His record with gold, money and women was not good and he eventually confessed to the murders adding that the troopers fired at him first. As one of the trooper's guns was still fully loaded and the other had but one shot fired from it, this was thought to be unlikely.

12. C. A. Bernays, *Queensland Politics during Sixty Years 1859-1919*, Gov. Printer, 1919 p355

It only took the jury an hour to come out with a 'guilty' verdict.[13] According to Warden W.R.O.Hill, Griffin was hanged in 'a dress suit, and lifted his long fair beard to let the hangman put the rope under his chin.' To the last, he maintained he was innocent. To try to prevent the grave robbing that was prevalent at that time Griffin was buried with the body of a second man, brought in dead from the *Tinonee,* in a coffin above him. However, this ploy did not deter the grave robbers. Griffin's body was decapitated and the grave restored. Hill states that the skull was still secreted away in Rockhampton in 1907. Following the sensationalism of the day, Hill reports that the rope allegedly used to hang Griffin was cut into short lengths and sold at a shilling a piece.

The following year at Rockhampton, Halligan's murders were tried and hanged. It was a good place to be out of. Palmer evaded the police successfully after he fled Rockhampton by hiding out for several months in caves and in the thick scrubs around Gympie. He was an excellent horseman with a superb thoroughbred horse and, in a chase, the police were soon left well behind. Palmer at times met, and shared a billy of tea with, a number of locals but was always on his horse and away at the first sign of danger.[14] At last, ill and tired of being a fugitive he decided to give himself up. There was a price on his head and he arranged secretly with a friend to tell the police where to find him and to collect the reward. This would then be split with Palmer's wife, a girl of seventeen.

Palmer was calmly waiting when the police rode out to get him. Had he wished he could have shot them on sight but he was weary to the point of exhaustion. There was controversy over who actually got the reward, the informant or the arresting police. W.R.O. Hill cites lawyer Wickey Stable as the informant. Zac Skyring suggests Palmer's wife herself, or her woman friend and is doubtful if the young wife saw any of the money. Hill states Stable received the reward and 'apportioned it as requested by Palmer' to his wife.

Palmer and Archibald confessed to the crime but Williams claimed his innocence. Williams and Palmer were hanged on the same day, the occasion of a violent thunderstorm. Hill wrote of the scene saying, 'Williams, who was a remarkable man, made an eloquent speech, and bitterly cursed his accusers while heavy rain fell and the lightning flashed, and the thunder rolled overhead. It was a tragical scene.'

While on the Gympie field Jack met the legendary prospector James Venture Mulligan discoverer of the Palmer, the Hodgkinson and other fields, a friendship

13. W. R. O. Hill, *Forty-Five Years' Experiences in North Queensland,* Pole Brisbane, 1907 p180
14. Zachariah Skyring, *Gympie in its Cradle Days,* Gympie & District Historical Society, 1917 and 1985 p65

which lasted until Mulligan's untimely death, the result of an injury incurred when he tried to stop a fight in the Mt. Molloy pub.

Mulligan tells of how he and Dr. Jack, 'a gold digger then', went into Billy Flynn's hotel in Gympie on their way to the dance room.[15] A scuffle broke out ahead of them when a burly man named Bluey accused a stranger of burning his finger with his cigar. The newcomer apologised if he had done so, saying that it was accidental. Bluey would accept neither the explanation nor the apology. He wanted to fight. The stranger was hesitant, suspecting – quite rightly – that it would not be a fair fight. Dr. Jack offered to second him and a ring was cleared in the centre of the crowded dance hall.

The women stood at the back on chairs and forms and an inner circle of men crouched on the floor. The stranger dropped Bluey with an early punch but immediately another man, Long Bill, Bluey's second, rushed at him. Hamilton called to fight fair and leapt defensively in front of the stranger. Not to be denied, Long Bill swung a left at Hamilton who dodged it and dispatched his own left with such force that 'the first part of Long Bill's anatomy to touch the floor was the back of his head.' Long Bill recovered in a quiet corner and the dancing continued.

Years later, on the Palmer, Hamilton told Mulligan the sequel to the episode. On his way back to his camp, the stranger emerged from the shadows and thanked Dr. Jack for saving him from the mob for he was sure it would have developed into a free for all fight. He was Detective Hanley and the young woman with whom he had so frequently danced that night leading up to the fight, was the wife of Palmer, one of the suspects in Halligan's murder. Through her, he was trying to find Palmer hoping to persuade him to give himself up.

Dr. Jack stepped in again on the side of the underdog when, one Sunday at the One Mile Gully at Gympie he fought Billy Scott, said to be the Professional Heavyweight Champion of the West Coast of the U.S.A.[16] Scott, in Gympie to seek his fortune on the goldfield, attacked a man in one of the shanties and his 'crowd' jumped into the fray to give him a hand. Hearing the melee, Jack arrived and without a thought for his own safety jumped into the milling bodies to give the lone man a hand. In the course of the brawl Dr. Jack was knocked down and kicked so badly one arm was out of action.

Honour wasn't to be satisfied until Scott and Hamilton, the American champion and the Gympie hero, had it out one-to-one the following day.

15. W.R.O. Hill *Forty-Five Years Experiences in North Queensland,* Pole. Brisbane, 1907 p133
16. *'Beachcomber' Heroic Queenslanders,* Sunday Mail, 31 August 1941

Hamilton, as the local man, was to select the site, a level, grassy patch where they could stage a 'most comfortable fight'. He did. The marathon bout lasted for an hour and twenty minutes and ended with Jack knocking out the West Coast Champion. Scott was carried to his hut with Dr. Jack staggering along behind. He treated Scott's injuries and nursed him back to health. By patching up a defeated foe, Jack made another devoted admirer. 'Beachcomber' was obviously another, saying Jack 'was a beloved, almost legendary figure, a knight without fear of reproach, a gallant enemy, and a great gentleman.'

A few years later Scott was in a Sydney hotel when word came through that John Hamilton, a.k.a. Dr. Jack was the newly-elected Member for Gympie in Queensland's Legislative Assembly. So pleased was Scott for his hero's success that he bought champagne for the bar and told the crowd that drank it the story of the fighting medico who beat him with only one hand and then nursed his battered body back to health. In his opinion, the tall, slightly-built Dr. Jack towered like a god among men. Fortunately for Jack, as his defence of the underdogs left little time for mining, let alone doctoring, other doctors came to the field.

THE CURTIS NUGGET OF GYMPIE

Excitement always ran high in Gympie's early days but it reached a new high when, taking over an abandoned claim near the Lady Mary P.C. on Sailor's Gully which ran into Nash's Gully, George Curtis and his nephew Valentine Brigg sunk their mine shaft a couple of feet deeper and unearthed a prodigious lump of gold. With it Curtis found an enterprise more profitable than his old job as a Maryborough sheep-scab inspector. The nugget took Curtis's name and from it 906 ounces of highgrade gold were recovered. The value of the Curtis Nugget was reputedly well over three thousand pounds. There was litigation over its ownership with the owners of the adjoining claim contesting the boundary. Fortunately the matter was settled reasonably amicably in the local Miners' Court with Dr. Jack as one of the presiding magistrates. It was hard for the owners of the Lady Mary P.C. as the huge nugget was found barely five feet from her abutment with the once-abandoned claim but they did well for themselves too. Out of one crushing of seven hundredweight of stone came 1488 ounces of gold.

The Curtis Nugget wasn't the biggest found in Australia. The Welcome Nugget found at Bakery Hill in Victoria ten years earlier weighed in at 2200 ounces and at Dunolly, also in Victoria, the mammoth Welcome Stranger turned out to be a hundred ounces heavier than that. But the Curtis Nugget had its admiring public and was put on display, firstly in Maryborough and then in Brisbane and Sydney with spectators being charged for the privilege. The money raised went towards building a hospital in Gympie and before long there were three new doctors in residence, Benson, Dowdney and Ryan. Jack could concentrate on mining.

He still looked after anyone who sought his help and quite a few continued to do so. Amongst the population was a fair proportion of Irish miners who'd brought their families with them and because of basic conditions and minimal hygeine, Jack was often called to treat the wives and children as well as injured miners. Mostly they were treated without payment, but some time later when Jack's fortunes were low, a delegation of Irish miners came to him with over three hundred pounds worth of gold and wanted him to accept it.[17] He was most reluctant. He didn't really wish to be tied down as a general practitioner and after all, he was a lone miner while they had family responsibilities. The Irish felt they had their country's honour at stake. They assured Jack that every pennyweight of gold had been contributed by an Irishman. It was all their own work. For a while, thinking of his hero's misfortunes in Mexico, Jack was tempted but finally he refused their gold. Once satisfied that it was more a matter of goldfield mateship than mere charity, the Irish families kept their offering and with it their high regard for the doctor in moleskins.

For some time the Gympie field remained in the hands of individual miners and their mates. The eminent geologists of the time scorned the idea that gold could be found there at any great depth from the surface. This, for the time being, discouraged big-time investors and companies from speculating on the reefs. Shafts did not need to go down to great depths so capital was not a huge consideration. The basic requirements seemed to be 'a little woodwork and a horse or two' and enough money to provide living expenses. Often men worked in pairs. One worked the partnership's mine while the mate worked for wages in another to keep them both going until that lucky strike was made. The horses were used to work the simple machinery, whips and whims, which were used in place of the motorised versions. Horses were popularly in use to operate the puddling machines and pumps, though at times, bullocks were used to pump water. A good draught horse could cost from twenty-eight to thirty-five pounds, a considerable amount when gold was worth between three to four pounds an ounce.

VOTE FOR TOZER AND HAMILTON: LABOUR AGAINST CAPITAL

Commissioners King and Charles Clarke held their Miners' Court at first in a bark hut on Commissioner's Hill. Later a slab Court House was built where

17. *'Beachcomber' Heroic Queenslanders*, Sunday Mail, 31 August 1941

differences could be settled and appeals attended to.[18] An early session was the scene of a musical protest when one of the more eccentric of the town's doctors, Dr. Byrne, known as the Jumping Doctor, led miners in a demonstration round and around the building while a controversial hearing was in progress, all the while singing rousing tunes like *Marching Through Georgia.* It is not known whether this influenced the magistrates in their decision but it provided a bit of light-hearted fun for the community.

With Local Mining Rules being the authority to follow, members were elected to the local Miners' Court. Dr. Jack was selected to contest the election with the town's popular lawyer, Horace (later Sir Horace) Tozer. Tozer was held in high regard. He was the soul of tact and an excellent diplomat. He had to be. Frequently, being the only lawyer in town for some time, he had to handle all cases and unless someone decided to represent himself in court, to serve both sides. Such were his skills in diplomacy that often both parties considered they'd done very well in the dispute. This earned him the sobriquet of the Miners' Friend and he capitalised upon it. With Jack he devised a hard-hitting slogan 'Vote for Tozer and Hamilton: Labour Against Capital: The Poor Man Against the Rich Man!'

It was a valuable exercise. Who could resist that egalitarian cry and be classed with the capitalists? Opponents who campaigned on the same platform were greeted with boos, hisses and sometimes missiles while the Miners' Friends got the cheers and the votes. The Miners' Friends win was decisive. They received 467 votes each against the opposition's 176 votes per man. Jack usually polled highly but he outdid his normal popularity later up North at Halpin's Creek and California Gully where the polling results were decidedly incredible.

Once the Court House was in use, Police Magistrates were appointed to hold the courts there. Some of the early magistrates like Sellheim, Lukin, Hodgkinson (who had the goldfield named for him) became distinguished figures in Queensland's mining history. Mowbray, who was Warden on the Hodgkinson soon after the field opened in 1876 was also an early Police Magistrate. Unfortunately Jack's relationship with him was not as cordial as it was with the others and on the Hodgkinson they clashed quite regularly.

With the bureaucracy set up, red tape looped-out in all directions to entrap the miners. It cost two shillings and six pence to register a claim and a similar sum to get a permit to cut posts and slabs for winches and to line the mines. Tax-wise nothing seems to have changed that much over the years.

18. William Clarke, *Reminiscences, Gympie Times,* 16 October 1917

Gympie had its Chinese miners as well as Europeans. They were allowed to work alluvial gold as 'puddlers' but only after the other miners had abandoned the workings. They were extremely industrious and painstaking and extracted all the gold it was possible to mine from these claims. Their method of digging a mineshaft differed from the European way. Chinese shafts were often but not always, round – perfect circles that rarely, if ever, caved in.[19] The other diggers used square- or rectangular-topped shafts and these seemed to lack the stability of the round holes. Rarely were the shafts, some fifteen to twenty feet in depth, filled in and wandering horses, cows and at times, humans fell in.

The Chinese also branched out to another vocation they performed with excellence – that of gardening. Abundant fresh fruit and vegetables complemented the plentiful supply of cheap beef from the surrounding stations and the wagonloads of other foodstuffs brought up from Brisbane. Lamentably the fruit, especially the watermelons, had to be protected from marauding children who regarded the Chinese and their gardens as fair game. In time schools were built which curtailed some of the raids and gave the young ruffians something more constructive to occupy their time. A Roman Catholic school opened on Calton Hill. One of the trustees was a young priest who came early to the field, Father Matthew Horan. He and Dr. Jack had a lot in common with their generosity and their altruistic outlook and became life-long friends, Father (later Dean) Horan reading the eulogy at Jack's funeral.

FATHER MATTHEW HORAN AND THE BABY

Father Horan loved to tell the story of his arrival in Gympie. When he could equip himself with a horse he always rode and was, also like Jack, an excellent horseman with a preference for thoroughbred breeding in his mounts. As he dismounted in Mary Street a group of miners drinking in one of the hotels opined it was "Time to leave."

And so they did, but not for long. There were very few groups which did not welcome Father Horan's inclusion in their company.

Father Horan's earthly possessions at that time were his horse, its saddle and bridle, his swag, a valise of church papers, altar cloths and vestments and a half-empty tuckerbag.[20] Before it was time to make camp that evening, a friendly miner leaving for Brisbane and not one of the Father's particular religious persuasion, gave him his small tent and some camping gear on the pretext that

19. Mrs. Oswin, *Gympie in its Cradle Days*, Gympie & District Historical Society, 1917 and 1985
20. Dean Horan, *A Record Pastorate, Gympie in its Cradle Days,* Gympie & District Historical Society, 1917 and 1985

he'd have no further use for them. By week's end, another benefactor, an Anglican, gave him a large tent to use as a church for his Sunday service and it remained in use until the first church was built.

Father Horan continued to live in his tiny tent on the hill. He liked the view, he said, but the creek was a fair distance away and that was where he had to go to do his washing. He'd been carefully washing his own clothes, the altar linen and vestments for a couple of Mondays when he noticed a group of 'motherly' women taking a great interest in his work and finding it quite mirth-provoking. Finally his ineptitude proved too much for them to bear. They came over.

"Here," said the spokeswoman, "we'll do that."

And this group of (Presbyterian) women cheerfully continued to do the Reverend Father's washing until he had not only a church and a presbytery but also a housekeeper to look after them.

While religion at Gympie was friendlier and more ecumenical than on many fields, sometimes national chauvinism was less tolerant. There were a large number of New Zealanders on the field, many of them of Irish descent and of Father Horan's flock. As they all stuck together, when one struck a lean patch, they all suffered although gold may have been abundant just across the way.

An article in the local newspaper inferred that gold was plentiful and that anyone with the slightest competence could make a tidy fortune. Only absolute duffers could fail to make a go of it. The New Zealanders – on a lean patch – took this to mean they were considered incompetent miners and reacted angrily. The newspaper wouldn't retract and the Kiwis went looking for trouble. They decided to pull down the newspaper building. It was by no means a substantial structure so the probability of them doing just that was high.

In vain, Father Horan tried to dissuade them with pleas to remember their Hibernian background and their faith. Turn the other cheek. But his entreaties went unheeded. The men attached strong ropes to strategic parts of the building.They lined-up like teams in an aggressive tug-of-war but all on the one side – and, just as their leader gave the order to pull and Father Horan asked the Lord to intercede, he spied the shining blade of a tomahawk leaning against the post to which the main rope was attached.

Trained to act on omens from Above, the leader had just got to "One! Two! Three! Heave!" when the Reverend father cut the rope, 'almost' as they said in the aforementioned newspaper 'unnoticed'. The revolutionaries collapsed in a heap and fortunately, when they sorted themselves out and picked themselves up, saw the humour of the situation and went home. If they knew the cause of their sudden downfall, they refrained from taking revenge. Matthew Horan remained one of the most popular men on the field.

Kindness and humour often go hand in hand. Though Father Horan usually rode when he went to Brisbane on church duties, there were times when he took the coach. On one occasion he was travelling in the box seat with the driver when a woman passenger was having trouble with two small children inside the coach. The swaying movement and the confined space of the usually crowded carriage often caused motion sickness and the smallest traveller was badly stricken. A word to the driver and the coach stopped while Father Horan asked could he take the ailing baby on the box seat with him. The mother, unable to cope, readily agreed and in a short while the baby was over its indisposition and having a frolicsome time on the boxseat with the two men so that they forgot there was a horse-change and a stop-over coming up.

Naturally, there was a group of miners in attendance, well-primed during their wait for the mail at the public house. The sight of the priest nursing the baby was cause for unbounded merriment but Father Horan merely shrugged and laughed it off with an enigmatic, “Just one of the flock.”

THE ABORIGINES OF GYMPIE AND ZAC SKYRING

Gympie being in a fertile valley with plentiful supplies of bush tucker, there was a large Aboriginal population. For the main, there was little friction between the old and the new residents though there were incidents on both sides where the behaviour could not be condoned. The district was also the home of the bunya pine and every third year was a bumper year for bush tucker – bunya nut time. At that time, tribal differences were put aside and neighbouring tribes were allowed to come and share the largesse. Mostly they camped out of town on old camp sites near water. Any who did venture too close to town were quickly driven away by miners or the police. Used to the socialistic life of their own camps they were often prone to seeking to ‘share’ the white man’s possessions without being invited to do so.

When the indigenous people met there was often fighting between the tribes and coalitions of allied tribes. The warfare was spirited and went on for some hours until someone was killed or the injuries mounted too high. Then one side would give way, peace was declared and a great time had by all fraternising at a corroboree. A civilised war. The corroborees were reflections of things that happened in their lives. The white men’s inexplicable habits often provided comic relief for the watchers as clever Aborigines mimicked their actions.

Sometimes, while men fought on one side of the hill or creek, the women battled it out on the other side. The rules were the same and the skirmish also ended amicably in another corroboree. Performed to the beat of the clickstick and the age-old wind instrument the didgeridoo, the corroborees resembled the white man's ballet.

The Skyring family was one of the numerous families who befriended the original inhabitants. Zac Skyring senior was reasonably fluent in several of the Brisbane dialects and was able to make himself understood by the local tribes when he moved to Gympie.[21] Both Zacs, senior and junior were given tribal names and learnt additional Mary Valley languages. These they collected to aid the survival of the ancient tongues. Similarly the Skyring girls learnt the language of the Mooloo tribe and attended women's corroborees. Some were purely for entertainment while others, like the one that warned of the dangers of the whiteman's rum and tobacco, acted as moral lessons. Later, while staying in Sydney with her daughter, the poet and writer Zora Cross, May Cross nee Skyring told some of Zora's friends of her experiences and demonstrated part of a corroboree she'd learnt in her youth. This is supposed to have given John Antil the inspiration to write his ballet 'Corroboree'.[22]

JOHNNIE CAMPBELL, HIGHWAY ROBBER AND RAPIST

There were undesirable types in both races. Johnnie Campbell was one. He was the victim of his association with whitemen who had little compassion for any man whatever the colour of his skin and who had very few moral principles. An excellent horseman who spoke fluent English as a result of his earlier liaisons and education, he had a pleasing personality and became a favourite with men and women alike, telling stories of bush life with dry humour. But for the two years he was on the run, he was feared by both races.[23]

Women especially lived in dread of seeing his shadow fall across the doorway of their shack or gunyah. Travelling mostly on foot and during the hours of darkness, he was usually accompanied by an Aboriginal woman, who had to keep up with his incredible speed of travel. Seen in one area today, he could be thirty miles away tomorrow. He knew all the hiding spots and short-cuts but eventually it was his own people who brought his reign to an end.

21. Zachariah Skyring, *Hunting with the Wide Bay Blacks, Gympie in its Cradle Days*, Gympie & District Historical Society, 1917 and 1985
22. T. Cross, Private papers
23. Zachariah Skyring, Op Cit

His last companion was an Aboriginal woman of Junoesque proportions, the biggest woman most people had seen. She seemed to be genuinely fond of him and did not desert him after he was captured even though it was she who unwittingly led to that capture. It was Johnnie's habit to send her ahead in to a camp to see if the people would accept him. He waited hidden on the outskirts ready to fly at the first sign of antagonism. She went into a camp at Noosa, near Gympie and reported to Johnnie that the people were friendily disposed. He came in but the fear he'd instigated by his rape and plunder – he never murdered – plus the incentive of a sizeable reward tipped the scales against him. He put up a valiant defence but there were just too many. He was overpowered, tied up 'with Mrs. Goodchap's hempen clothesline' and held until the police took him into custody.

He was tried in Ipswich and sentenced to fourteen years for highway robbery. The women he raped were too frightened to speak against him until one man induced his daughter, one of Johnnie's victims, to come forward to give evidence. After this he was found guilty of rape, the penalty for which was death by hanging. The Kabi tribe were gathered and taken to Brisbane to witness the punishment. Johnnie's body was de-capitated, as was often the case in those days, in the interests of phrenology and the head sent to London. There is no record of the findings of the examination for criminal traits.

More community-minded was Billy Beaston,[24] Aboriginal groom for Captain Beaston the Clerk of Petty Sessions at Gympie. From Beaston, Billy learnt quite a lot including very fluent and persuasive English. At every conceivable opportunity Billy would mount his political platform and address any who gathered out of curiosity on his projected candidature for the Legislative Assembly. As he rightly pointed out, the Aborigines had no say in the running of the country and, as the natural owners they had a right to do so. At the end of his campaigning Billy took around the hat for donations that would allow him to try for a seat in Parliament but, as one of his contempories, Zac Skyring wrote, he 'never got to the plush benches of the supreme council of the state.'

24. Zachariah Skyring, *Hunting with the Wide Bay Blacks, Gympie in its Cradle Days,* Gympie & District Historical Society, 1917 and 1985

3

NORTH TO THE PALMER 1870

Gympie's floods of 1870s. Gold production declines. Macrossan leaves for new field at Ravenswood. Jack and many others go north to the Palmer. Jack sets up hospital. Swam flooded river to rescue Paddy Shanahan. Takes-up the Queen of the North P.C.

GYMPIE'S BIG CAKE AND THE SOCIAL LIFE

For a lengthy period Gympie prospered. From its inception until the mid-seventies 600,000 ounces of gold were mined. Gold was found in the traditional quartz and one particularly spectacular reef was called the 'Jeweller's Shop'. Quartz and gold were in almost equal quantities along its sinuous length and the effect on the viewer was breathtaking. In the first full year, the gold escort safely conveyed over 80,000 ounces of gold to the banks' vaults returning tax fees of 2170 pounds to a gleeful Queensland Government.

Of the rival mining companies, the Wilmot Extended[25] had perhaps the most lavishly stocked jeweller's shop. Their crushings were phenomenal during the first five years of reefing. The dividends paid to shareholders were rivalled only by those paid to investors in Mt. Morgan's golden mountain. Monkland 7 and 8 were Wilmot Extended's toughest competition. One crushing of 212 tons yielded almost 4000 ounces of gold – nearly nineteen ounces to the ton. These big crushings saw the gold escort fee dropped after much agitation by the mine

25. *Gympie in its Cradle Days,* Gympie & District Historical Society, 1917 and 1985

owners, from sixpence an ounce to four pence half-penny but after one particularly good crushing the owners calculated the gold fee of one hundred and forty-nine pounds was excessive and hired a special coach. For forty pounds, it would take the gold to Brisbane.The gold was in retorted form and in one solid ingot called the Big Cake. On the first night from Gympie the coach pulled up at Cobb's Camp as the men escorting the gold became drowsy. They were dozing off after a hearty supper when one guard suddenly remembered the unprotected gold. The guards rushed out expecting the worst but fortunately the unattended twenty thousand pounds treasure was intact and unharmed in the coachbed. At three hundredweight, two quarters and nine and a half pounds avoirdupois (about 203 kgs) it was safe from lone highwaymen unequipped with hydraulic hoists or sophisticated lifting devices.

When the Big Cake arrived at its destination it couldn't be weighed on any of the existing gold scales because of its size. These scales weighed in troy weight, the normal measure for gold (twelve ounces troy to the pound as against sixteen avoirdupois ounces) so its official weight had to be recorded in avoirdupois, the measure used for more mundane things like flour, sugar and potatoes. It was also put on display at sixpence a look and the money raised was given to charity.

After the first crushing plants arrived early in 1868[26] there were prodigious crushings. To see if the two batteries of five heads apiece, each driven by a twelve horsepower steam engine, were working efficiently 0.3 ton of stone was put through on a trial run. The stampers must've been doing the job reasonably well as 0.15 ton from 1 North Caledonian yielded over 450 ounces of gold and a similar parcel from the Caledonian P.C., 367 ounces. A phenomenal 817 ounces were extracted from just 0.3 ton which, if the process could have been similarly repeated for the full ton of stone, would have produced a massive 2,723 ounces to the ton.

Tent city gradually gave way to wooden buildings with shingled roofs. Timber, including red cedar and pine, was plentiful and the local artisans produced slabs, milled and pit-sawn timber for building as well as transporting logs to the metropolis. Often the river was used to raft the big cedar logs to the port at Maryborough and in the floods of the seventies, even before the disastrous floods that followed in the nineties, enormous losses were suffered. In the later floods logs were found beached at Noosa and the estimated cost of reclaiming them led to their being abandoned there.[27]

26. *Gympie in its Cradle Days,* Gympie & District Historical Society, 1917 and 1985

In the interim between canvas and timber, builders used bark extensively and quite a profitable industry evolved to produce the commodity, something like a sheet of textured masonite. The Aborigines were past masters at this art of removing cylinders of bark from the trees and flattening them to a convenient sheet over a fire. Some of the more entrepreneurial became industry leaders and did very well out of the project.[28]

Most Aborigines were confined to camps outside the towns – where they were not welcomed without a good excuse – or on cattle properties. The indigenous population was usually blamed – rightly or wrongly – for any pilfering that took place and 'smoking out' expeditions moved these people farther out from the town limits. Gunyahs were burnt. The men were resilient and soon rebuilt their homes on a new site while the women, 'carrying babies or puppies', fossicked through the ruins for anything of use. Deaths occurred from diptheria, measles and other diseases brought in by the whites and for which the Aborigines had no resistance. Sexually transmitted diseases brought sterility and death to them as well.

In town, cow-dung fires were kept burning at night to deter the mosquitoes which carried the many fevers and agues but fever was still rife and the doctors were kept busy at the small public and smaller private hospitals. With the other doctors on the field Dr. Jack spent more time at his mine. His calls out to women in labour and children with diptheria or croup decreased, except from the loyal Hibernians who trusted him above all other doctors. Miners still came to him readily to have gaping cuts stitched and broken bones splinted.

Rainwater tanks fashioned from the zinc linings of the crates in which drapers imported their wares took away some of the need for drinking the often polluted river water and cases of dysentery eased. As the town gained permanence, clubs were established and throve. Old newspapers tell how thousands rolled up to enjoy the athletic sports meetings that were held regularly. A Hunt Club was inaugurated where mounted locals chased kangaroos with all the fanfare of an English foxhunt and a kennel of kangaroo dog 'fox hounds' was kept at the Nine Mile under the supervision of Houndsman Sauer. Fresh meat was sent down by coach almost every day to ensure the dogs enjoyed the correct diet.

Lawyer Wicky Stables, Father Horan and Jack Hamilton were keen huntsmen and always well-mounted. Some of the more adventurous horsewomen,

---

27. Thomas H. Steele, *When the River was Jammed with Logs, Gympie in its Cradle Days,* Gympie & District Historical Society, 1917 and 1985
28. Walter G. Ambrose, *Chinese and Aborigines, Gympie in its Cradle Days,* Gympie & District Historical Society, 1917 and 1985

including the elder Skyring girls, joined in the fun of the chase riding sidesaddle. They re-appeared that evening, dressed in their finest satins and laces for the hunt ball that followed. The Gun Club met at O'Leary's Crossing for their pigeon shoot and at Duramboi for duck-shooting. As a marksman Dr. Jack struck keen competition from publican Captain Croker but he managed to keep his reputation as a marksman intact and placed betting winnings in the pockets of his followers.

A race club held irregular meetings throughout the year but held two two-day carnivals each year without fail on the Widgee Crossing road. Money was plentiful and wagers and prize money were high. Horses came to compete from as far afield as N.S.W.

Almost the entire population attended the meetings, rigged out in their best, to cheer their fancies on. The booths, which did a roaring trade, served their purpose though they were like many of the structures at latterday bush picnic race meetings, saplings supporting bough or canvas roofs. Travelling musical and variety troupes performed at the township's theatre, the Varieties.[29] Again, the whole town turned out to enjoy the presentation. Miss Jo Geogenheim's dramatic company staged plays regularly for over twelve months and Billy Barlow was a popular singer. A troupe of Gympie amateurs presented *The Bohemian Girl* to crowded houses at the Varieties and group of musicians regularly entertained. Among the more favoured instruments were jew's harps, timbrels and concertinas.

The floods of March 1870, though nowhere near as devastating as the ones yet to come, ruined many mines near the river, creeks and gullies. Some of the miners were getting itchy feet and others, ever ready to seek the quick profits alluvial gold could offer had already left for the Cape, Ravenswood and the booming Charters Towers field. Townsville, after languishing in Bowen's shadow for so long, came to life with these rushes and Townsville businessmen were eager to promote further exploration that would bring more business to their town. Their promotion program was persuasive. More miners went north.

Other professions went too. Wickey Stables made the change to Ravenswood but his time there was brief. Here, in a violent thunderstorm he took refuge in the cottage of Warden Hackett only to be struck dead by lightning within minutes of reaching what he thought was sanctuary. Stables was killed outright, Hackett rendered unconscious and the housekeeper who vowed she'd been struck too,

29. Walter G. Ambrose, *Chinese and Aborigines, Gympie in its Cradle Days,* Gympie & District Historical Society, 1917 and 1985

died some weeks after the event. All this is reported by W.R.O. Hill in his *Forty Five Years' Experiences in North Queensland.*[30] He stayed with Warden Hackett but was absent the day the ill-fated Stables selected Hill's chair at his host's table. Stable's widow later married Robert Lord, M.L.A. for Gympie and after his death married a third time to become Lady Horace Tozer.

Hill later went south to Gayndah where he 'held a multitude of billets.' This was fairly normal for a Police Magistrate in a rural town. He lists his official duties as 1. Police Magistrate 2. Gold Warden 3.Clerk of Petty Sessions 4.Acting Land Commissioner 5.Land Agent 6. Registrar of the Small Debts Court 7. District Registrar 8. Registrar of Births, Deaths and Marriages 9. Savings Bank Officer 10. Commissioner for Affidavits 11. Agent for the Curator of Intestate Estates 12.Electoral Registrar and 13. High Bailiff.

In his spare time he acted as 1. Churchwarden 2. Superintendent of the Sunday School 3. Choirmaster 4.President or Secretary of most of the charitable and sporting institutions.[31]

JOHN MURTAGH MACROSSAN

Dr. Jack's miner from Ridgelands also went north. Jack the Hatter, John Macrossan. Macrossan, slightly built, with a stoop and a look of perpetual ill-health despite his piercingly clear blue eyes, made his way to Ravenswood where he worked, characteristically on his own, the Saratoga mine. A quiet and a solitary man he was nevertheless incensed by the way the unpopular Warden Hackett was treating the miners. He came out of his seclusion to organise them into the Ravenswood Miners Association. His resentment of the treatment meted out to the miners and the libel Hackett encouraged that Macrossan was a de-frocked Jesuit priest, came to a head. He horse-whipped the Warden in the 'muddy bed of Elphinstone Creek before the bridge was built' as Hill reports in his memoirs.[32] Hill 'capsized Macrossan into the sludge', rescued his landlord and sent for the police.

The local justices called for a ridiculously high thousand pounds bail requisite. Miraculously it was immediately forthcoming, not from Macrossan nor from his miners but from 'a notorious shanty-keeper named Annie Smith' who was lifted onto miners' shoulders and carried around town by an admiring crowd. Macrossan was brought before the District Court in Townsville and fined thirty pounds. A disapproving Hill remarked that 'it made for him a name very popular

30. W. R. O. Hill, *Forty-Five Years Experiences in North Queensland,* Pole Brisbane, 1907 p56
31. W. R. O. Hill, Op Cit, p101
32. W. R. O. Hill, Op Cit, p55

among a certain class of miners'. It may well have done so but it also made the name of John Murtagh Macrossan very well known in other circles. He was in politics 'for two of the stormiest decades of Queensland's political history' as Secretary for Public Works and Mines, Colonial Secretary and Secretary for Mines.[33]

He never once forgot his miners and the north. He was in Queensland politics for almost twenty years and played a vital role in the series of pre-Federation conferences. In all that time he was still a loner – 'feared, respected and for the most part trusted. He was little loved, perhaps, but he was surprisingly little hated.' So Harrison Bryan wrote of him in his biography of Macrossan. What politician could ask for more?

JAMES VENTURE MULLIGAN AND ROBERT PHILP

James Venture Mulligan, ever one to seek the gold at the end of the rainbow, tramping over lonely rough miles to locate it but loath to settle down to the everyday mining of it, had already left the banks of the Mary River. Writing of Mulligan, Robert Logan Jack penned, 'a prospector is an explorer in every sense of the word (one is tempted to add 'only more so'). Mulligan was the Crown Prince, if not the King of Prospectors and a loyal friend.'[34]

Jack Hamilton met Robert Philp as a lad at a Gympie cricket match. Here Philp was introduced to merchant James Burns who employed the youth and later made him a partner in the firm known and respected all over the state – Burns Philp. Philp was younger than Dr. Jack but they had a similar Scottish upbringing and many interests in common. Philp was not as interested in mining as Jack Hamilton was and it's told that after Philp lost money in early investments on the Etheridge field at Georgetown he was offered a half-share in the Day Dawn mine at Charters Towers. That investment almost guaranteed to put him well back in the black. The price asked was one hundred and twenty pounds but Philp's burnt fingers were still stinging. He refused the offer and spent the money on a horse and buggy, an expensive purchase as the Day Dawn produced well over a million pounds in dividends by 1903.

THE GOLD RUSH TO THE PALMER

The rushes to Cape River, Ravenswood, Charters Towers, the Gilbert and the Etheridge lured away hundreds of goldseekers from Gympie but when the

33. Harrison Bryan, *Macrossan, Jack the Hatter Queensland Political Portraits,* Murphy and Joyce U.Q.P. Brisbane, 1978
34. Robert Logan Jack, *Northmost Australia Vols. 1 and 2,* Robertson, Melbourne, 1922

Palmer's wealth became known many businessmen left too. The Palmer, like every other goldfield, provided little but bitter disappointment and disillusion to those who hurried there hoping to pick bucketfuls of gold nuggets from the ground and to return within an exceedingly brief period as millionaires to a life of ease. Many did make fortunes. Many lost what money they originally had, tried to get work or left the district. A carrier named Stockden took his German wagon up to Cooktown by boat and did very well carting to the Palmer. Stockden was an unusual man in that he drove his eleven horses in reins while seated on the wagon like a coachman. Teamsters usually walked beside their teams.[35]

Another teamster, Jack Hamill, was an excellent violinist and carried his instrument everywhere he went. When things went wrong – as they often did with bullock and horse teams – he was in the habit of playing his fiddle 'to soothe the beasts'. It worked. His fellow teamsters sang and jigged about while both men and beasts quickly acquired a happier, more positive frame of mind and the difficulties were soon resolved.

Some mine managers and small businessmen from Gympie also successfully made the changeover to the new field, including one of Gympie's foremost butchers Arthur John Fisher and his son A.J. jnr.[36] The steamer was used in the timber trade and was rather small to be making the long trip up the coast to Cooktown but it carried thirty men and twenty horses disembarking them safely at the Endeavour River just seven days later. At Cooktown the Fishers bought horses, packed them with gear and provisions and walked another one hundred and twenty miles to the Palmer, carrying their swags across their backs.[37] Hundreds were doing the same things but others were also travelling from the opposite direction, sick, disillusioned and broke after failing to make their fortunes on the River of Gold.

With many of his friends already in the north and more ready to go, Jack Hamilton joined the rush to the Palmer. As Gympie settled to its role of a prosperous country township rather than a rambunctious mining camp Jack had given some thought to his place there. At one stage he was fired-up to go to New Zealand to help in the Maori Wars but the trouble ended. A Maori even gained a seat in Parliament and life over there seemed to indicate plain sailing ahead. With his background in mining and the Gympie Miners' Court he was approached by Arthur Macalister, three times Premier of Queensland and once Secretary for

35. *Gympie in its Cradle Days*, Gympie & District Historical Society, 1917 and 1985
36. A. J. Fisher, *Bark Hut and Canvas Shanty, Gympie in its Cradle Days*, Gympie & District Historical Society, 1917 and 1985
37. A. J. Fisher, Op Cit

Public Works and Goldfields, to take on the Commissioner's position on the Palmer.Hamilton declined the honour, preferring as a contemporary writer recorded, 'the use of a miner's pick as a means of livelihood'.Macalister's nickname was Slippery Mac owing to a predilection for swapping sides in politics and not being terribly careful about keeping promises.Maybe Jack, with his preference for honesty, simplicity and straight dealing, was a little doubtful about the outcome. He didn't accept. He caught a boat to Cookstown – later to become Cooktown and was among the earlier arrivals on the Palmer.

WILLIAM HANN AND THE TREE KANGAROO

The outstanding bushman and cattleman, William Hann, pioneer of *Maryvale* station in the basalt north of Charters Towers was instrumental in starting the Palmer rush. He was sent north by the Queensland Government to inspect Cape York Peninsula. Kennedy's papers were lost after his tragic death and the young Jardines kept mostly to the western side of the Peninsula on their way to Somerset. Little was known of the eastern fall and the Government was eager to find out what type of country lay there. They were still keen on developing the colony from the north down with Somerset already established as a base at the tip.

Everyone was on the lookout for gold and Hann promised a half-pound of tobacco to any of his party who could find the elusive mineral. The honour (and the tobacco) went to Surveyor Frederick Warner who panned for and found gold on the Palmer. As Hann had to overcome difficulties in that wild terrain that would have stopped less experienced and determined men he was hesitant to publicise Warner's finding and start a rush. What Warner had found was gold, but not in payable quantity. Hann was concerned it might prove a fizzer and considering the rough terrain and the isolation, a rush there could bring in its wake hardship and even death to would-be gold seekers.Not only was the country itself rugged and inhospitable but it was far removed from the closest white settlements.Another member of Hann's party was Thomas Tate. He was a survivor of the ill-fated brig *Maria* where because of his medical training he was employed as Acting Surgeon. On Hann's expedition he was the botanist. Taylor was the geologist. Stewart, Nation and Jerry an Aboriginal made up the party.Jerry's bush skills were especially appreciated by the party and its leader.[38]

Hann missed out on two discoveries that trip. As well as not recognising the extent of the goldfield and losing the honour of being the finder of the River of Gold, he lost the kudos of a second discovery made with the help of his station

38. Robert Logan Jack, *Northmost Australia Vols. 1 and 2,* Robertson, Melbourne, 1922 p374

'boy', Jerry, an Islander. Jerry showed Hann and the botanist, Tate, marks in trees which he said were made by kangaroos that lived in the tree-tops. Not doubting Jerry's words, Hann kept his eyes open and saw these strange kangaroos too but when he reported the sighting, the idea of kangaroos bounding about from limb to limb in the canopy was so incongruous no one believed him. It took Carl Lumholtz nearly a decade later to convince the scientists that Dendrolagus Lumholtzii really did exist.

MULLIGAN AND DALRYMPLE ON THE PALMER

James Venture Mulligan thought Warner's gold find was worth a second look. If a cattleman – or a surveyor – could find gold, surely an experienced prospector could do better. At Warner's discovery site Mulligan and his party found little more than Warner did. They found 'colours', gold too insignificant to be weighed, but they found payable gold just about everywhere else along the river. Mulligan's wasn't the only party that set out to improve on Warner's find but he was the first back to break the news.

Three months after he left the Etheridge, Mulligan and his mates returned to Georgetown and a wire from the telegraph station there soon alerted Cardwell and beyond that 'Prospectors Mulligan, Brown, Dowdall, A. Watson and D. Robertson got 105 ounces on the Palmer River which they prospected for twenty miles. They say nothing of the country outside the river. Nearly all are leaving here.' It was September 3rd 1873.[39]

Miners quickly quitted Gilberton which had been constantly harassed by native warriors angry at being dispossessed of their hunting grounds. It was also beset by tropical fevers. It was here that Dalrymple caught the fever that contributed to his early death. Miners left in droves from the Etheridge and with Mulligan as their guide flocked to the Palmer. From all points of the compass they came, from southern Queensland and from the southern colonies, from New Zealand, America and, of course, from China.

To find a suitable port for the Palmer Dalrymple was instructed to lead a Government expedition from Cardwell sailing north. He immediately rode to Georgetown and conferred with Mulligan and his mate Alec Watson on the best possible site before returning to Cardwell where the leaky vessel the *Flying Fish* was waiting. The Georgetown ride was a long, arduous one done in the shortest possible time with Dalrymple still suffering the effects of his fever and from a bad fall from a horse. Cook's landing on the Endeavour River was the most suitable

39. Robert Logan Jack, *Northmost Australia Vols. 1 and 2,* Robertson, Melbourne, 1922 p417

site. Dalrymple had hardly berthed there and was drying out his boat when a steamer was seen entering the heads. It was the *Leichardt* carrying nearly a hundred optimistic diggers, 'New Zealanders, Victorians, New South Welshmen and Queenslanders' as well as the Warden, Howard St. George and MacMillan, the surveyor. They'd needed no help from the Government to find a port. Captain Cook's choice a century ago was good enough for them.

The official party had horses and the services of Jerry of tree-climbing kangaroo fame but the diggers were on foot. MacMillan wanted to spell the horses to allow them to recover from the voyage before they set off inland but five miners, impatient at the delay, left ahead of the party. Only one survived.

WILLIE WEBB AND MATES EAT 'BANGO'

The story of the trip out is told by Willie Webb who lived out his later years in what became Cooktown. At the wide Normanby River, which was where the party of five split up and four went permanently missing, shots were fired at some Aborigines.

'I do not know why,' Webb wrote, 'as they had not interfered with us.'

MacMillan and St. George fired over the heads of any marauding tribesmen when necessary to frighten them off but 'some of the men fired straight at the blacks'.Like many of his time, Willie Webb abhorred the senseless killing. The place of the attack was named Battle Camp and an inquiry was held into the shootings but no charges were laid.Unhappily it set the scene for many subsequent deaths on both sides and in 1879 Dr. Jack brought up in Parliament the case of a young Aboriginal woman killed near Battle Camp when an attempt to abduct her, present her with gifts hoping she'd give a good report to the tribe on her return, went abysmally wrong. The woman was killed.The slaughter continued.

The miners were heavily laden with swags and supplies, averaging about seventy pounds apiece. Some even tried to carry heavier loads but as the party progressed much was reluctantly jettisoned. The Government party took the lead and blazed a tree-line for the others to follow. 'The Queenslanders,' wrote Webb proudly, 'were keeping up with the leaders.'

MacMillan 'discovered' another sizeable river and named it the Laura after his wife. Here poor Hann missed out for the third time. He'd already named it the Hearn for his wife's family.

When the party ran short of food, Willie Webb and his mate Tom Lynett went off in search of fish in Pine Tree Creek. Willie took his kangaroo dog as well as the fishing lines. They caught nine fish and the dog landed a whopping big 'roo, so heavy they had trouble carrying it back to camp. But they managed and Webb 'never saw such willing and enthusiastic co-operation as there was in skinning, dressing and cooking the meat.' There were over a hundred men in the camp and all had a 'bit'. The dog was given the bones.

The party crossed the range into the Palmer fall at the Conglomerate and were relieved to see horse-tracks in the river sand. They knew they were close to their destination. On 13th November MacMillan rode out, found diggers already at work and returned with Inspector Dyson from the Etheridge. The following day they moved on to the main camp of 'a few hundred diggers' at Palmerville.

Tucker supplies were very low. Webb and his mates had been reduced to eating 'bango' – a gruel of boiled sugar and flour – and what they could catch. There was a five-horse dray at the camp with a few bags of weevily flour left – at two shillings and sixpence a pound. The teamster's wife, a Mrs. Neil, was according to Willie Webb 'mounted on the dray and conducted the sale of the flour. If the lady didn't like the looks of you, or found fault with your manners or thought she could read in your eye any question as to whether the battered pannikin she measured with really held a pound of flour – you went without.'

Jack Edwards and Alf Trevethan, butchers, were there with cattle. Only nine were left of the sizeable mob they'd put together and soon they were sold. There was no salt to preserve the meat for storage so the diggers sundried it during the day and smoked it over the fire at night. All were pleased to have arrived. The trip from the banks of the Endeavour took about seventeen days.

CORFIELD THE CARRIER CROSSES THE MITCHELL

Jack Hamilton was already at Palmerville when W.H.Corfield,[40] then a carrier but later a parliamentary ally of Jack's, arrived there at the end of 1873 before the Wet season set in. Corfield had travelled from Ravenswood and Charters Towers, through the Etheridge and the Gilbert. He and his mates had a 'very exciting trip passing Fossilbrook, Mt. Surprise and Firth's stations, crossing the Lynd, Tate, Walsh and Mitchell Rivers. These were all running strongly from the storm rain that normally precedes the monsoonal Wet.' At the Mitchell there were forty teams 'of all sorts and sizes waiting to cross.'

Corfield's mate considered the river had receded far enough to attempt a crossing the following day and they led the way over. Everyone crossed safely and there was 'quite a little village of people of both sexes' camped that night on the

40. W. H. Corfield, *Reminiscences of Queensland 1862-1899,* Pole Brisbane, 1920 p53

north side of the stream. From there the travelling was 'easy' for the next thirty miles that took them to the Palmer.

With forty teams arriving at the same time the supply of goods for once exceeded the demand. Corfield's mate, Wilson, left with one team to purchase conjointly with Corfield another load, while Corfield sold his bullocks and horses and set up a store under a tarpaulin to sell the goods left with him. A new diggings called Purdie's camp started some forty miles upstream so, hiring a team, Corfield carted his merchandise up there and quickly disposed of the lot at very high prices. A two hundred pound bag of flour brought twenty pounds, two shillings a pound, a little cheaper than Mrs. Neil's and without the weevils.

At Purdie's camp a packer, Frith, had a horse cast a shoe and rather than miss out on the packing, the owner offered to pay their weight in gold for five horseshoe nails to replace the lost shoe. He found a seller almost immediately in one Billy Yates, a carrier.[41] The nails were put on one side of the scales and gold added to the other pan until the weights were even. The horse normally carried a hundred and fifty pounds of loading which at a shilling a pound returned a gross of seven pounds ten shillings from which the cost of the nails had to be deducted.

Corfield was amused when a man who came in to T.Q.Jones' store to buy a needle objected to the price – one shilling.

Jones was stung to retort, 'Good God man, look at the price of freight!'

Corfield tried to estimate the price per ton involved in the carting if just one needle – plus freight – cost a shilling.

By the time Wilson returned with the loading, Corfield had obtained new teams and was engaged to take four tons of supplies to Palmerville. It was on this trip he met Dr. Jack and wrote, 'Among the Palmer diggers, Hamilton was extremely popular because of his prowess as an athlete and his medical ability which was given gratuitously to all. He was said to have been concerned in some of the many South American revolutions but although we were friendly from this time until his death, he never alluded to such an occurrence. I realised, however, that he was very reticent as to his early life and the gossip may have had some foundation.'

With Jack coming back to Australia as an assisted immigrant on the *Wansfell* at the age of twenty-one it doesn't leave much time between his medical studies and his emigration for partaking in revolutions.Perhaps his support for Juarez in Mexico was re-located further to the south.

41. W. H. Corfield, *Reminiscences of Queensland 1862-1899*, Pole Brisbane, 1920 p50

During one Wet season on the Palmer, Corfield was caught in the rain. He had with him as spare boy, a young Aboriginal lad he'd reared from early childhood. The boy caught a cold and despite Corfield dosing him with whatever medicine he had available, he developed pneumonia and died.[42] Corfield never left him during his illness, nursing him faithfully day and night and was devastated by his demise. 'He was a splendid little fellow and I missed him greatly.'

An experienced bushman, Corfield, when with his wagons, habitually made his camp some distance away while he made what looked like a dummy bed under the wagon. That was the proven way to dodge most of the spears thrown in the night. The first mailman to the Palmer, Johnnie Hogsflesch, had horses speared and was finally speared himself, though thankfully, not fatally, in the course of his mailman's duties. Not so fortunate was the boatman stationed at the Laura River to ferry travellers across. He was speared and reportedly eaten as well.

DR JACK TREATS 3,000 CASES OF FEVER & SPEAR WOUNDS

Meanwhile Jack's time was taken up with his mining activities and his medical work. A Government paper on deaths at the Palmer from October 1873 to July 1875 gives the main cause of death as dysentery, seventy-four deaths of the two hundred studied. Fever accounted for a further forty-seven and other medical causes, thirty-eight. Nine were drowned, two murdered, one committed suicide and spear wounds, snakebite and other acts of misadventure counted for the remaining twenty-nine.

On one occasion when Jack was camped at German Bar, a man came urgently seeking help. His mate, Paddy Shanahan had been speared at a spot about thirty miles downstream. As Shanahan's mate was close to exhaustion, Jack told him to rest while he packed a medical kit. That done they started back to the wounded man. It was dark by the time they reached the river and by the sound of the water it had risen considerably since Shanahan's mate crossed it. The mate, so soon after his long ride, wasn't keen on re-tracing it in pitch darkness. Jack assured him he knew the way and could even take a shortcut which would reduce the distance somewhat. He was willing to go alone and rode his horse into the floodwaters.

Almost immediately it was struck by a floating log and horse and rider parted company midstream. Both returned unharmed to the bank they'd just left. Here Jack instructed the mate to take care of the horse, put his oilskin-wrapped medical kit in his shirt, tied it to the top of his head by the shirt sleeves and successfully tackled the river again on his own. Once across he made his way to Palmerville

42. W. H. Corfield, *Reminiscences of Queensland 1862-1899,* Pole Brisbane, 1920 p53

where he woke his friend Isaac Butler and explained the situation. Without a second thought, though by this time it was midnight and there were still ten miles to go, Butler accompanied Jack to the wounded miner's camp.

Shanahan was speared in the arm and the exceedingly swollen limb was causing almost unbearable pain. Jack removed the embedded spearhead, stitched the wound and applied a healing ointment from his kit. The grateful man offered Jack his entire gold supply, the product of much hard work, three ounces but Jack refused to take it. He saw his action as merely observing the rules of common decency to a fellow miner. Shanahan claimed, possibly with a vestige of truth, that Dr. Jack's selfless exploit saved his life.

Years later, another man named Shanahan – M.W. Shanahan - passed on a message to Dr. Jack about a wondrous El Dorado in the furtherest north of the Peninsula.[43] Jack did go up there on several occasions and secured a Government prospecting grant to work that country. His friend John Dickie found payable gold in the Coen area possibly with the assistance of that grant. His message from Shanahan concerned a strike that went by the name of Dead Man's Gold as each party finding it succumbed but for one lone survivor. This man upon reaching civilisation would put together another party only to have a repeat performance. The Deadman's Secret was bequeathed to Dr. Jack with gratitude and charts by the final survivor. Shanahan had some fun with the charts. 'Behold then Doctor Jack pouring over those bewildering charts, pouring the whole flood of his analytical intellect on the variations shown by two apparently contradictory charts; but the discrepancy is only seeming, and is entirely owing to the compass by which the chart was made out always pointing south-south–east of the camp, and the other chart's compass always indicating precisely five hours and thirteen minutes beyond the fourth day of next week.'

Jack did go up the Peninsula soon after on Mining Commission business but didn't report finding the Dead Man's Secret mine. Shanahan recounted that too, adding a weak pun. 'The Doctor cried Eureka! a couple of years back, and shot through the end of Cape York like a comet, but he still didn't come it over the secret, which remains a secret still.'

Shanahan and Jack weren't the only ones who expected gold to be found further north and Shanahan thought that if he, Jack Hamilton 'offered the secret for acceptance to the first man who would venture a sudden and mysterious death as a probable concomitant, there would be a big rush for the first place.'

43. M. W. Shanahan, *With the Cape York Prospecting Party,* 1896

Dr. Jack's popularity on the Palmer never waned. It is estimated that while he was there he treated about three thousand cases – fever, dysentery, spear wounds and injuries due to mining accidents and drunken brawls. If his patient had no funds it did not matter, Jack treated him just the same but if someone had the gold to pay and did not – watch out! Jack was a very different man with his temper raised.

One man who fancied himself as a bare-knuckle fighter decided to take his chances and abscond without paying. Jack let it go believing the man would pay when he struck payable gold but found the New Zealander was already working a rich claim in Snider Gully and could well afford the fee. Passing the gully one day, Jack asked the man if the debt had slipped his mind. The digger said that it hadn't, he'd give Jack an ounce of gold and if Jack wanted more – the debt was nearer to six ounces – he could 'knock it out of him.'

The ex-patient was a good three stone heavier than his doctor, big and strongly built. Dr. Jack was tall and slender but with muscles like whipcord. The two peeled off their shirts and, as Harry Harbord, one of the crowd who watched, recounted, 'in three rounds the medico cut his opponent to pieces, so he not only had to pay up with the five ounces of gold he owed Hamilton but he again temporarily became Hamilton's patient.'

Harbord himself was nursed back to health by Dr. Jack when he had fever and 'shakes'. It took seven weeks to get him back on his feet during which time Jack acted as nurse and cook as well as doctor, treating the fever, washing the patient and his sweat-soaked clothes and feeding him. The diet was usually a rice gruel stewed in an iron pot over the campfire. The cost of treatment and board was nine pounds a week or three ounces of gold.

Harbord also confirms Jack's prowess with a revolver and tells with parochial pride how the Palmer champion beat Colonel Withans, American champion pistol shot in a contest and that, added Harbord appreciatively, 'was in the days of the Wild West.'

Jack later built a small shack hospital in Maytown to treat his patients. Near him, across the creek, was the only other doctor at the time, a German Dr. A.H.F.R. Korteum. Korteum came to Queensland and went to Charters Towers just after the Franco-Prussian war. He then followed, like so many others, the goldrush to the Palmer. With patients paying in golddust he did well and later moved in to a more civilised practice in Cooktown where he also held the position of German vice-consul. Both doctors had their own graveyards at the Palmer with a 'few hundred lying between the two of them', according to Harry Harbord.

Even then there were doubts as to whether Dr. Jack was a 'registered' doctor. But Palmer diggers like Harry Harbord were prepared to overlook minor details when Dr. Jack was so willing and able to attend to wounds and broken bones and was inordinately 'clever at treating fever, ague and dysentery, the principal illnesses on the Palmer.'

That, after all, is what really counts.

1. Hann's 1872 expedition. Thomas Tate seated left. Norman Taylor seated right. William Hann, back row left and beside him the dicoverer of gold on the Palmer, Surveyor Fred Warner. - *R. L. Jack*
2. Dalrymple's 1859 Expedition. Sitting - Stone (left), Dalrymple (right). Standing - Sellheim (left) and Henry. - *Bowen Historical Society*

1 | 2

1. Gold Escort.
2. Whim horse hauling ore from mine. - *Richard Daintree*

1

2

1. Miners' camp, Cape River. - *Richard Daintree*
2. Mary Watson. - *Queensland Museum*

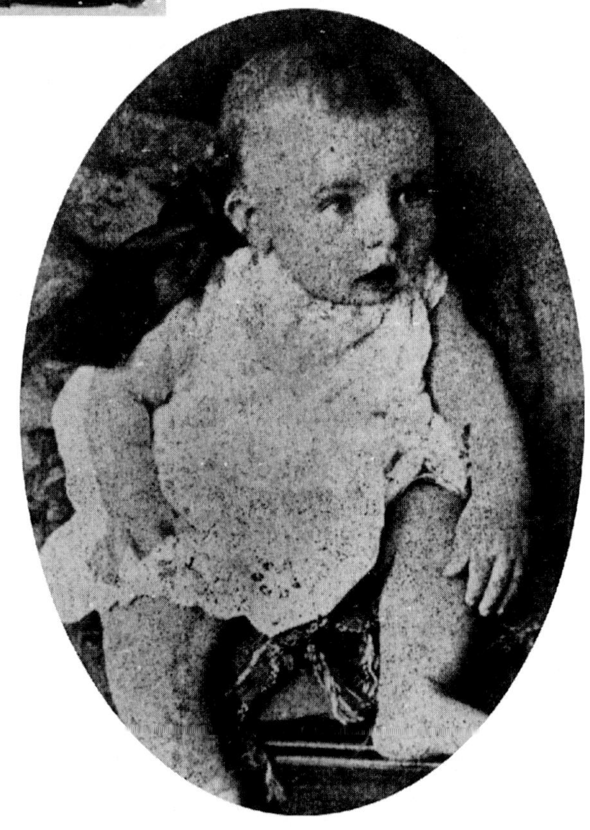

1 2

1. Tank and paddles used in the escape from Lizard Island. - *Queensland Museum*
2. Baby Ferrier Watson. - *Jillian Robertson*

1

2

1. Mounted Police setting off on patrol from Cooktown. - *John Oxley Library*
2. Artist's impression of Chinese landing at Cooktown. - *Illustrated Australian News 1874-5*

1

2

1. Maytown coach ready to leave from Cooktown Hotel (later Commercial Hotel). - *Queensland Heritage*
2. Christie Palmerston. - *Eacham Historical Society*

1 2

1. Machinery at the site of the Queen of the North. - *Peter Bell*
2. Two Wenlock miners with their gold takings for a week. The gold was stolen in Laura. - *Ted Youngman*

1. Robert Christison of Lammermoor, jockey, after winning a double in 1857. - *M. M. Bennett*
2. Frank and Alick Jardine about the time of their trek. - *Byerley*

1. Sluicing for gold. - *Glenville Pike*
2. Grave of the Prince of Prospectors, Mt. Molloy.

1. Robert Philp, passionate Notherner and Queensland Premier. - *John Oxley Library*
2. Thomas McIlwraith, transcontinental railway visionary and State Premier. - *John Oxley Library*

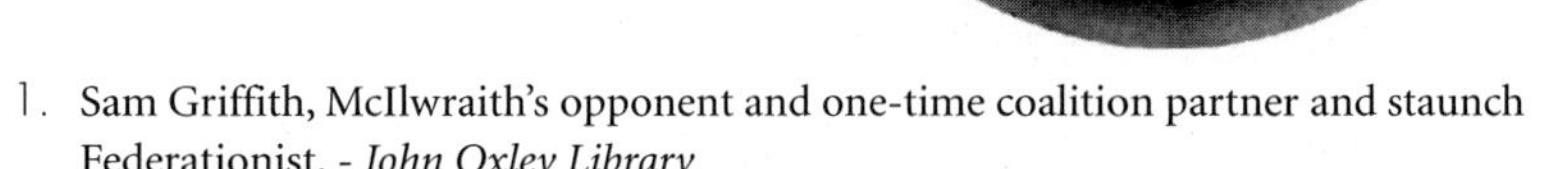

1. Sam Griffith, McIlwraith's opponent and one-time coalition partner and staunch Federationist. - *John Oxley Library*
2. John Murtagh Macrossan (or, as he spelt it, MaCrossan), Jack the Hatter, ardent Separationist and leader of the Northern Nine. - *John Oxley Library*

1. James Nash, discoverer of Gympie's gold and saviour of the infant colony. - *Gympie Times*
2. Maytown identities of 1875. 'Dr.' Jack Hamilton (with bottle), Dr. Ahearne, Jack Edwards and another resident. - *John Oxley Library*

1 1. Poole Island Meatworks. - *Illustrated History of Queensland*

2 2. Warden W.R.O. Hill and Native Troopers. - *W.R.O. Hill*

1

2

1. Sluicing for gold. - *Illustrated History of Queensland*
2. Miners panning-out. - *Illustrated History of Queensland*

1. Street scene, Brisbane floods of 1893. - *Courier Mail*
2. Griffin, Gold Commissioner and Warden, Rockhampton and murderer of Troopers Powell and Cahill of the Gold Escort. - *W.R.O. Hill*
3. Palmer, Queensland bushranger and co-murderer of the goldbuyer Halligan. - *W.R.O. Hill*

1

2

1. Pearling schooner *Sagitta* sank with almost all the pearling fleet at Bathurst Bay during cyclone Mahina 1899. - *Illustrated History of Queensland*
2. Army training camp Enoggera, World War 1. - *John Oxley Library*

4

THE NEW RUSH TO THE HODGKINSON 1876

Warden St. George appointed. Chinese flock to the diggings. Warden Coward at Byerstown. Inspector Douglas blazes a shortcut track through Hell's Gates. Mulligan discovers the Hodgkinson. Miners leave for Hodgkinson in 1876. W.R.O. Hill relieving warden at Hodgkinson.

THE CHINESE MINERS ON THE PALMER AND IN LUKINVILLE

The Palmer wasn't an easy goldfield although even the rawest of new chums had the chance to find some alluvial in the gullies. Some claimed you could pull the grass up from the creek-side, shake the roots clean and find gold in the discarded dirt. Others were lucky enough to pick up nuggets straight from the stony beds of the streams. Crevice miners found nuggets in cracks in riverbed rocks but isolation and lack of reliable roads compounded the usual difficulties of life far from a coastal town.

Mulligan tried to warn the would-be diggers against the perils of travelling during and immediately preceding the Wet season. There were plenty of wide, swift-running and dangerous rivers to cross and hosts of travellers found themselves stuck between swollen watercourses, or bogged down in the tea-tree country with dwindling, almost non-existent tucker supplies.Often they were surrounded by Aborigines angrily resenting the intrusion into their lands. But of course, few listened.

Gradually wagon roads were made for the teamsters, some of the tracks with their fine stone-pitching are still in evidence today. Sandy crossings and boggy stretches were 'corduroyed'– paved with saplings laid side by side across the road like ribs in a corduroy fabric. Corfield and his mate corduroyed one section of sand to get across the Normanby to the Deighton. While he and his mate laboured in the sun, felling saplings, trimming, carrying and laying them in position, some teamsters drinking rum in the shade jeered at them and made disparaging remarks on their workmanship.

Corfield had the last laugh. He and his off-sider crossed safely and to the unbelieving chagrin of their tormentors burnt the saplings before they left. It was a difficult job as the timber was still green but it afforded them considerable satisfaction. Normally bushmen would eagerly lend a hand for the common good but Corfield was furious at the teamsters' behaviour and was determined they would not profit by their rudeness.

Prices escalated once the horde of diggers arrived. Any sort of a horse brought sixty pounds and beef on the Palmer was sold for a shilling a pound, a fourfold increase on Gympie's threepence. There were, as yet, no local cattle producers and the butchers allowed a good margin for droving expenses. Contrarily, sea freight from Brisbane and Sydney to Cooktown came down in price.

Howard St. George was Warden when the first Chinese arrived. He called a meeting to ask what the miners wished him to do. The Warden reminded them that once the Chinese were on the field he was responsible for the welfare of all miners, Chinese and European alike.[44] The easy pickings had gone and most of the men at the meeting were ones who hadn't done well. The losers were tired of the hard life, the scarcity of food, the ever-watchful Aborigines and the lack of profit.? 'It's only fit for Chinese,' was the meeting's verdict. 'Let them have it.' The Chinamen flocked in by the thousands. It's estimated between twenty and thirty thousand passed through Cooktown en route to the Palmer.

The Chinese didn't always sell their gold through the normal channels. It was worth much more in China and large quantities were smuggled out as boats came stealthily in to the secluded bays along the coast north of Cooktown. Rumour has it that one Chinese headman, through his bonded miners, smuggled out an average of a thousand ounces of gold a month for four years. Cape Flattery, home of the Foxtail palm, was a popular base for Chinese going to or coming from China and who didn't want to be bothered with Cooktown Customs' red tape.

44. G. Bolton, *A Thousand Miles Away,* Jacarandah Press, Brisbane, 1963 p54

Apart from jealousy, that would've occurred no matter what race was involved, when some do exceedingly well and others barely make tucker, the main argument against the Chinese was their sense of hygiene. Water was all-important in a land where doctors and medicines were scarce and dysentery rife. By mutual consent the miners agreed to leave the smaller running streams for drinking water. The Chinese didn't see the need quite so clearly and used even these streams for bathing and washing clothes and utensils. Cholera and typhoid were introduced and the bacilli are still evident in some streams.

New roads were opened and boatmen waited at major rivers to ferry travellers across. Warden Coward's road to Byerstown where he presided, branched off MacMillan's Road after the little Oakey and ran south-west by way of Kings Plains lagoons, across Ted Hahl's crossing on the Normanby where Ted kept a boat in readiness It then more or less followed the present Mulligan Highway up the Byerstown Range, down Spear Creek to Byerstown. Maytown was further west.

Inspector Douglas's track greatly shortened MacMillan's original road by branching off before Battle Camp, crossing the divide between the Normanby and the Laura rivers, following Quartz Creek to its head and crossing to the Palmer fall through the narrow Hell's Gates defile. This was the main track used by packers, horsemen and those travelling on foot. Wagons still had to go the long way round. Hell's Gates was wide enough only for single file traffic. All roads were dangerous for wayfarers but Hell's Gates, the ideal set-up for an ambush, was the worst. By choice, travellers went that way in groups for safety reasons. Many are the tales of people who fell victims to the Merkens' spears and were, more times than not, eaten.

Maytown, as the principal settlement, grew with mushroom speed. Early in its history, three banks, the New South Wales, Australian Joint Stock and the Queensland National put up corrugated-iron roofed buildings to safeguard their gold. Their managers were often required to ride out to do business and in contemporary photos looked more like stockmen than pen-pushers in their broad hats, leather belts, moleskin pants and top boots. Hotels or shanties were everywhere and tubs of champagne regularly provided drinks for all when a miner struck it rich. There were two newspapers, Boyett's *The Golden Age* and Gibson's *Palmer Chronicle.* The town even boasted a School of Arts with an extensive library. There was no Christian church but the Chinese had their joss houses.

By 1876 there were a dozen stores with double that number of establishments run by Chinese shopkeepers. There was animosity between the two races of shopkeepers as well as between miners. The Chinese merchants used coolies to bring their goods out from the port of Cooktown, each man carrying almost as much on his bamboo pole as could be put on a packhorse. They worked in strings of fifty to two or three hundred and were a very cheap form of transport working for little more than subsistence rice while the other storekeepers had to pay packers and teamsters a very high price to carry their loading.

Many Chinese died on the Palmer and were buried there if found by their compatriots or had their bones sent back in earthenware urns for burial in their homeland – accompanied often, it is said, by many ounces of illicit gold. In China, wealthy merchants saw a foolproof way of increasing their riches. Holding families as hostages, they sent the able-bodied men as their bondsmen to the Palmer to win gold for the hierarchy at the most minimal cost and discomfort to themselves. The hapless Chinese coolies were often ambushed and the last man in a long line of basket-carriers was easy game. The Aborigines, it was rumoured, preferred the flesh of the Orientals to that of the Europeans which they claimed were too salty for their taste. Perhaps, someone suggested, this was a piece of propaganda hopefully circulated by the non-Asiatics.

Often, through ceaselessly carrying heavy loads, the coolies suffered a paralysis of the legs which prevented them from following their jog-trotting mates and were left to die of starvation or to receive a speedier death at the hands of the indigenes. But the Aborigines weren't the only ones to prey on the Chinamen. If a spear were found in a corpse the Merkens automatically were blamed and at times the real murderer escaped – with the victim's hard-won gold. The Chinese were poorly armed and travelling single file in long lines through the bush were prime targets for both black and white assailants.

On the credit side, the Chinese did the miners a great service producing tobacco, fresh fruit and vegetables, bamboo for buildings and furniture. They also provided, amongst other services, laundry and gambling amenities, to say nothing of the opium which, regrettably, was used to some to keep the 'tamed' Aborigines in a euphoric state of submission.

As the alluvial diminished, reefing took over along the main river and prospectors scattered to the gullies farther out and over the divide to the headwaters of the Normanby. In the very early stages gold could be garnered like 'grains of wheat' or simply picked up as small nuggets known as 'bar gold' in the crevices of the river's rocky bars. In places, the bars across the Palmer's bed were not unlike the 'ripples' put to trap gold in a sluicebox. Water carried the gold from its source and, because of its weight, simply deposited it in the rock fissures.

Oakey Creek, a tributary coming in from the south of the Palmer was one of the best prospects. In June 1874 the township of Kingston was set up and the Chinese moved in to work the creeks near Jessop's Gully and Stonyville. Jack Hamilton made the move to Oakey Creek although his 'town block' was still under his name in Maytown in an 1882 survey. He was given the chance to purchase a 'chemist shop' on the Oakey from a friend Thomas Ingham who was leaving for Cooktown. Ingham was one of the survivors of the brig *Maria* which was wrecked on Bramble Cay near Cardwell. Ingham had a shipmate from the *Maria* who also figured in the history of the Palmer and Hodgkinson goldfields. He was Tate who had been a doctor on the ill-fated *Maria* but who later took up the study of flora and fauna and was William Hann's botanist on his trip north. The Tate River, a tributary of the Lynd, itself a branch of the Mitchell, bears his name. Not surprisingly, Dr. Jack jumped at the offer to buy the chemist shop and its contents. The price was a pound of gold.

In 1876, with a party consisting of George Marshall, John Meyer and William Duncan Larsons, Dr. Jack also applied for the lease of a mine, the Queen of the North 1[45] at Gregory's Gully on the road to Maytown just east-south-east of the town. It was on the Mountain Maid Reef, one of the first to be worked on the Palmer and one of the best producers. The discoverers, Bremer, Meirer and Brede,[46] found extremely rich specimens just under the surface in 1874. The workings at the eastern end were relatively shallow with the reef dipping to the south-west. On the adjacent Gregory Gully, Chinese worked lucrative alluvial and were credited with several nuggets weighing 80 ounces in the year Jack and his partners bought the Queen.

The discoverers sold out to Hewitt and party and in February 1876 Jack applied for the lease. This was granted just before Christmas in 1877. Water was a recurring problem with the level at about 250 feet down but the mine was 'stoped', the auriferous veins removed, to the water level. Early crushings were very profitable and yielded four to six ounces per ton of stone. The production came from a reef approximately the size of the Ida, Comet and Louisa,[47] three other notable producers and the yields were comparable. It was worked on three levels until by 1884 the supply of the yellow metal all but ran out.

While all this was going on Dr. Jack still had his hospital tent on a ridge to the southside of Oakey Creek. One of his ex-patients wrote, 'if you had a dose of fever

45. D. W. De Havelland, *Gold and Ghosts V4,* Australian Print Group, 1989 p488
46. D. W. De Havelland, Op Cit, p488
47. D. W. De Havelland, Op Cit, p492

and ague – or 'shakes' as they called it on the Palmer – your mate would take you in to one or other of the doctors, rig up your tent or fly, make a bit of a stretcher and leave you there.' A comfortable bed could be fashioned with tripods of forky sticks at the head and foot to which were added a couple of sackbags with sapling rails passed through them to make a mattress with a bit of 'give' and up above ground level. The food was rice or rice gruel. 'You generally could not eat anyway. All you wanted to do was drink, preferably something sour.' Maybe the good doctor found plentiful lime juice, the cure for scurvy, in his newly purchased pharmacy. The cure could take a week or even a month or two. Jack sometimes varied the rice gruel diet with a stew or broth from kangaroo tail or squatter pigeon but though it gave him an excuse to practise his marksmanship, a meat diet was not considered suitable for fever patients.

Relations between the white miners and the incoming Chinese became grave. White miners maintained their claims were 'jumped' by the Chinese should they leave them even for the briefest period. The Chinese claimed they were beaten, ill-treated, humiliated, their pig-tails chopped off, while all the while they were innocent of any wrong-doing.

For many men it was time to leave. In 1875, the golden year, officially over 250,000 ounces of gold were mined. That fell to 200,000 the following year, still a very good return, but it dropped more drastically in following years. Mulligan located the Hodgkinson field. Miners were rushing to be the first there when it opened in mid-1876 a few months after Jack and his party applied for the lease of the Queen. But it was to be a reefer's delight rather than an alluvial miner's dream and many of the early travellers met equal numbers returning to Byerstown at the headwaters of the Palmer.

Upstream from Maytown, gold was also discovered at Lukinville, named for the Under-Secretary for Mines, George Lukin. This held at least the Chinese miners for some time but it became the scene of a 'war' between two Chinese factions – about 6000 Cantonese and a third of that number of Macao men. Warden Sellheim reported in 1878 'I regret to have to refer to some serious riots that took place amongst the Chinese at the beginning of the rush, during which four men were shot dead and many others were more or less seriously injured.'

There were probably hundreds of casualties, some survivors carrying very serious injuries. The 'clash of the tribes' as Sellheim referred to it was between traditional enemies. Macao, an island at the mouth of the Canton River was under Portuguese jurisdiction, just as, up till 1997, Hong Kong residents were 'British'. On both sides there were old carbines and Snider rifles but most of the weapons were the tools they used daily – axes, shovels, sticks, picks, iron bars and knives. Warden Sellheim tried to stop the fighting but with the numbers involved, once one skirmish was put down, another would flare. Many were killed in the fighting and of the hundreds wounded, large numbers died of their injuries or from infections. None the less, Cooktown merchants were very happy to hear of the rush itself even though the greater proportion of the miners were Chinese. It meant more business for them as well as the Chinese storekeepers. As a contemporary wrote, 'Cooktown sat up and smiled.'

Cooktown, on the coast, was at that time the only outlet for the Hodgkinson as well as the Palmer. There were many miles of most inhospitable country in between. None the less, with the Hodgkinson open, reefers deserted both the Etheridge and the Palmer in their eagerness to be where the action was. Townships sprang up everywhere and Thornborough became the capital for the satellite settlements of Woodville, Stuart Town, Union Camp, Kingsborough and others.

MULLIGAN DISCOVERS THE HODGKINSON & LEVELS HIS RIFLE AT THE RIOTERS

Mulligan discovered the Hodgkinson field on a Government-funded expedition from Cooktown, which though it was the port for the Palmer still had access to the old roads that brought in the first miners and teamsters from the Townsville hinterland. In Mulligan's party were his old mates Abelsen and Dowdell, Frederick Warner, Government surveyor and the man who earned Hann's half-pound of tobacco reward, Will Harvey, Jack Moran and an Aborigine named Charlie.

Leaving after the Wet, at the end of April 1875, they returned to Cooktown in late September, the Government grant exhausted and no gold of any worth discovered. Mulligan was itching for a second trip but no more Government money was forthcoming. He was sure gold was there – but where? Convinced one more trip would find it, he financed himself into another prospecting foray, taking Abelsen and Warner. Before Warner left he wrote to the Minister for Works saying he was stranded in Cooktown with no money to return to Brisbane, none to pay for paper and other necessities to chart where they'd already been and that he was going back with Mulligan. He could have added, 'So there!' but restrained himself. Miners, however, were most critical of the Government's unsympathetic and miserly action. While Mulligan and party were away, Emu Creek near Byerstown was opened up and it was Byerstown's chance to flourish.

Skilled bushmen that he and his party were, Mulligan managed to cross the flooded Mitchell in January and found some gold between there and the Hodgkinson. Unknown to him another party was also out prospecting – Macleod, Sefton, Kennedy and Williams.[48] They made each other's acquaintance in what could have been tragic circumstances. Hearing horsebells a short distance from their camp and curious to see who owned the horses, Abelsen went over. It was that time of day that comes closest in North Queensland to twilight. In the poor light Kennedy mistook Abelsen for an Aborigine about to spear one of the horses and fired at him.

Fortunately, for Kennedy was noted as an excellent marksman, the failing light saved Abelsen. The bullet went wide. Hearing the shot, Mulligan and Warner hurried over and the incident had a happy ending. The two parties, finding gold but still prospecting to reinforce their decision, luckily kept in touch, for a grass fire started up and an unusually strong wind swept it through Mulligan's camp, jumping his customary firebreak. Clothes, swags, rations were all burnt. Fortune stayed with them as their papers and the ammunition were both spared. Macleod's party happily outfitted Mulligan with the basics to get him and his mates back to Cooktown to announce their gold find. Should the Government relent and offer a reward, they'd share it.

At Byerstown, Coward was fishing for information and doggedly questioned Mulligan repeatedly about his trip until Mulligan succumbed, told him what they'd found and asked the Warden to promise not to tell anyone of the find until the party reached Cooktown and notified the proper authorities. But when they entered Cooktown they were greeted with newspaper headlines announcing the new gold strike. They were furious. The same evening that Mulligan reluctantly confided in Coward, the Warden had sent a mounted policeman posthaste to Cooktown with the news. Coward was never popular with the miners despite his pioneering of the track that bore his name. Complaints were made, petitions drawn up and signed. Coward was replaced by W.R.O.Hill in April 1876.

While in Cooktown Mulligan seized the opportunity to take out prospecting claims for himself and the others on reefs the parties had considered worth investigation. They were all agreed it was a reefer's field. That limited it to men with some money or backing to pay for the development of the mine. The little alluvial there, Mulligan advised was fit 'only for Chinese' who meticulously worked every inch of the ground and literally left no stone unturned.

48. G. Pike, *On the Trail of Gold,* Watson Ferguson, Brisbane, 1998 p58

On 23rd March 1876 a wire sent from Cooktown told of the party's find and predicted 'There will naturally be a perfect stampede of miners to the new El Dorado.' This was exactly what the prudent Mulligan wished to avoid. His reply was cautious. It would, he admitted, be one of the largest reefing districts in the Colonies but as for alluvial workings, he warned of disappointment rather than elation. In his own goldmining past he'd been through it, the desperation of hoping against all reason to find something just to keep going, fill the tucker bags and make life a little less forbidding. He understood too well what could lie ahead. People rushed in to Cooktown from their declining claims expecting Mulligan to lead them straightway to a new Paradise. He was extremely loath to take the responsibility but the rush was even greater than the original one to the Palmer.

Finally two hundred pounds were subscribed and Mulligan was persuaded – against his better judgment – to lead an advance party. He meant to go ahead with a select few marking a tree-line for the rest to follow when the rivers subsided. Instead he was followed by hundreds upon hundreds of ill-equipped and naïve hopefuls who struggled over bog and flooded rivers behind him. Mulligan left on 30th March, one of the wetter months. More than two thousand diggers were stopped by the flooded Mitchell. Some returned to Byerstown, got on the grog and ran riot.

One particular night in which dogs barked incessantly, above the usual roar of the merry-makers, Warden Hill was awoken by a shrill scream.[49] Getting his policeboys to follow him he hurried from tent to tent until he found a young woman on the ground in a pool of blood. Her right arm was severed above the elbow. The woman knew who did it – a jealous lover – but despite the flooded rivers he was never seen again. Miraculously, the woman recovered, thanks to Dr. Jack's ministrations and her own strong constitution.

Hill rode to the Hodgkinson in his capacity as Warden and reported that the alluvial rush must be stopped at all costs. Starvation and riots would ensue. As an encouraging postscript he added, 'The reefs are really good.' He may well have tried to stop the tide. While the prognosis for reefing was universally held to be good, there was no permanent warden – no law at all – on the Hodgkinson. The closest warden's offices were in Maytown, Byerstown and Cooktown, miles distant from the Hodgkinson and with the mighty Mitchell and its northern tributaries in between. An alluvial miner needs only himself and very basic equipment to get his gold but the reefer requires expensive machinery and capital

49. W. R. O. Hill, *Forty-Five Years' Experiences in North Queensland,* Pole Brisbane, 1907

to work his claim. None were keen to bring machinery in when there was no warden to legalise their claim.

In his brief two-day visit Warden Hill granted a number of reef protection areas which, some disgruntled miners alleged, tied-up too much auriferous country for a comparative few. The miners resented Hill but more so they resented the brevity of his stay. Kangaroo courts flourished. A horse-thief was flogged and chased from the field, his belongings confiscated. Petulant miners, dejected at finding the gold wasn't as easily available as the Palmer's 'grains of wheat', threatened to lynch Mulligan whom they unjustly blamed for their troubles. On one occasion Mulligan saved himself only by levelling his rifle at the ringleader. Thankfully, as the mob dropped back from their retreating leader and as Mulligan sat there on his horse in the middle of the angry mob and wondering what to do, someone had a change of heart. 'We're with you, Jim!' a man called from the thronging miners. A few cheers went up. The crowd dispersed and it was all over.

'Very few decent miners took part in the threats,' Mulligan explained to his supporters. 'They could not help being there in the crowd.'

There were other nasty incidents against Mulligan and finally Warden Howard St. George arranged for him to address a meeting at a Cooktown hotel to clear up the matter. Mulligan was given the chance to speak and he reiterated that he warned it was only a reefing field. He'd never encouraged alluvial miners to try their luck. On the contrary, he'd tried to deter them

Even the most dissatisfied had to agree with him. That cleared the air and soon after, St. George tactfully suggested that Mulligan might cease wearing a revolver on his hip during his stay in town.

5

THE QUEENSLAND PARLIAMENT & DR JACK HAMILTON, MLA FOR GYMPIE 1878

Jack moves to Hodgkinson.Takes doctor's position at hospital.Mary Ann Lovett's murder. In Brisbane Macalister resigns and Thorn forms new ministry.Macrossan in Parliament. Douglas takes over from Thorn.W.M.Mowbray warden at Hodgkinson. Jack on the committee to find a road to the coast. Trinity Bay (Cairns) the preferred port. Jack is returned as M.L.A. for Gympie. McIlwraith promotes rail line to Gulf. Jardine brothers overland to Somerset.

JACK AND THE MURDER OF MARY ANN SMITH

Leaving the Queen of the North in his partners' hands Jack, restless after hearing Jim Mulligan's reports, rolled his swag and left for the Hodgkinson. He arrived there without having to ford any swollen rivers in May 1876. As usual there was no doctor available and he had again to divide his time between being a miner, a medico and at times, a Justice of the Peace. It was a rowdy field. A woman was decapitated, a man murdered, an Aborigine disembowelled with a blow from a nulla nulla. It came as no great surprise when a miner came to Jack in a distressed state to tell him of a woman he'd found dead, her left arm severed above the wrist, the bones bare to the elbow. Her nose was also missing and one eye hung loose from its socket.[50]

50. *Cooktown Herald,* June 3 & 9 1877

McSweeney came across the body when he rode out on the 25th May to look for some missing bullocks. It was lying face-down sprawled on the ground, with the legs exposed to the knee. He didn't get off his horse. He could see from there that the woman was very dead and from his saddle he quickly looked for any obvious signs of injury or struggle before riding back posthaste for Thornborough. A party accompanied him back again – George Thompson, Gibson, Adams, Norton and a few other residents of the township. They turned the corpse over and found it was the body of Mary Ann Smith who had been living with a miner, Sam Lovett, who was also known as Lover or Doon.

Decomposition badly disfigured the face. The hair was separated from the scalp. In a macabre touch, her sleeve was rolled down to cover the stump of the severed wrist and the bared bones of the forearm. Lovett came along while the men were there and it was he, at the men's request, who moved the sleeve to reveal the grisly stump. They covered the body with boughs as some sort of protection and McSweeney was sent to fetch Dr. Jack their popular and respected miner, medico and magistrate. In the absence of the police and a warden his knowledge was indispensable. From their experience of him at Gympie and on the Palmer they had no need to look further afield. He was the obvious choice.

When they arrived and the men removed the protecting boughs, McSweeney was missing. The dark stain was indeed blood, suggesting haemorrhage as the cause of death. The severing of the hand would have been cause enough. Though they searched the surrounding area there was no sign of the missing hands. Dr. Jack considered the arm bones had been broken or hacked off. Part of the inner surface of both thighs was discoloured and one knee skinned as if injured in a fall. Dr. Jack took the head (on a shovel) to be washed in a nearby creek for examination. It was in such a very bad state of decomposition that injuries to the brain or to the windpipe could not be correctly gauged. There was a visible fracture of the upper part of the left jawbone such as could have been made by a blow to the head.

As the body had bled profusely, Dr. Jack ruled out the chance of a heart attack or of the left hand having been severed after death. Strangulation was a possibility but it was very hard to tell with the decomposition of the neck tissues. If the amputation had not caused the death, blood would flow freely for a short time after life was extinct but would soon cease. Jack considered that the amputation had been done with an axe or possibly by a powerful blow with a stick. The hands had definitely not been gnawed off by a predatory animal.

In his capacity as a Justice of the Peace, Jack Hamilton presided over a Magisterial inquiry into the woman's death after the body was found. At the preliminary hearing emotions ran high. Onlookers took sides and the procedure became lost in a near riot. With a few well-aimed punches aimed at the ringleaders, Dr. Jack quietened the loudest protestors after which the others observed a polite silence and allowed him to proceed with the inquiry.

Lovett was taken into custody for the murder but for some reason concerning state witnesses, the case wasn't heard for a year although another Circuit Court was held on the Hodgkinson during the intervening months.[51] William Leaning a.k.a. Yankee Jack or Big Jack, a Byerstown miner briefly in the Macleod's Creek area and who was seen with the woman on what was probably the night of the murder, was an early suspect. Some people linked Mary Ann's death with the case of the Byerstown woman whose hand had been severed and pointed their fingers at the Byerstown miner, Yankee Jack, and considered Sam Lovett was shamefully treated.

At Justice Sheppard's trial neither the judge nor the jury agreed with Lovett's supporters. Witnesses concurred that both Lovett and Mary Ann had been drinking heavily in the township and left for home in the company of the less-inebriated Yankee Jack. Lovett and Mary Ann were extremely intoxicated and kept falling down along the uneven track but that didn't stop Yankee Jack from propositioning Mary Ann. She acquiesced and it wasn't all that difficult to elude the drunken Lovett and to slip off to 'have connection' down the slope from the track though Lovett did follow and catch them up after the hasty intrigue was over.

People at the camp and a group of passing miners heard the sound of a fight, much yelling and loud screams. A woman's screams. It was a clear night and sounds travelled easily. Lovett was seen and heard chasing Yankee Jack, yelling and making threats. Yankee Jack took off back to the township to Dodsworth's camp where a drinking party was still in progress. He didn't join in but sat outside by himself 'in the dark'. Next day he left the settlement saying he was going back to try a prospect further up the creek. Later he was taken into custody but was discharged almost immediately.

The day after the camp heard the disturbance on the hill, Lovett appeared without Mary Ann saying she'd gone off with another man. He was throwing her and her belongings out. They'd been on good gold, Lovett told a miner witness. Mary Ann had gold on her and she could probably be murdered for it. But things

51. *Cooktown Herald,* June 3 & 9 1877

didn't add up in Lovett's story. Mary Ann's hat was found not far from the camp she shared with Lovett and drag marks were seen leading from there to where her body was found. No one could suggest any reason for the severed hands though someone remarked the natives were troublesome and could have been involved!

Lovett was said to have confessed to the murder to a miner friend and decided to face a trial rather than a Hodgkinson lynch mob. He had a long time to wait for the fair trial for which he opted. Justice Sheppard summed up to the jury at great length at 4pm on the last day of the trial. The jury went out a little after six and returned with a verdict of guilty before ten that night. They added a recommendation for mercy. The prisoner was asked if he had anything to say why the sentence should not be passed. He had not.

'His Honor, after a few remarks, with emotion, sentenced the prisoner to death.'

Dr Jack Defends His Medical and Personal Honour

One other long-outstanding mystery was cleared up that day as well, Dr. Jack's legal right to his title. In his deposition he stated with no attempt to prevaricate and without any prompting that he was a 'medical practitioner but not legally qualified under the Medical Act; had learned the medical profession for four years, during which time he had studied surgery and anatomy; had practised the profession of a surgeon for the last eight years; during that time assisted at surgical operations; had performed surgical operations of various descriptions; had amputated limbs from a living body; during the eight years had considerable experience and during the last three and a half years on the Palmer had treated over three thousand cases in that time.'

The miners subscribed to have a hospital set up and Dr. Jack was put in charge with the right also to a private practice. In the meantime, as the Special Correspondent from the Hodgkinson wrote in the *Cooktown Herald*,[52] 'We have as yet no warden, no mails, no police and having tried the experiment so long think we can easily do without them… The smash in the Ministry effects us in no manner; the former one did us no good, and we expect nothing from this one.'

The 'smash' referred to Macalister resigning in June 1876 after a no-confidence motion initiated by Macrossan. Following this George Thorn (for whom Thornborough was named) formed a new ministry, making use of a member from the Upper House for the first time in the Parliament's history. In reality, Griffith was the strong man of the Parliament and Thorn lasted only nine months

52. *Cooktown Herald*, June 28 1876

before retiring. The Douglas administration took power in the following April. Brisbane was then, as now, a whole world away from the Hodgkinson.

But they did finally send a warden before the reefers with the machinery they wanted to bring in from the Etheridge and the Palmer gave up in frustration. W.M.Mowbray was only in his mid-twenties when he was appointed. He was educated at home by his father, a scholar who gave private tuition in Brisbane. At eighteen young Mowbray began work in the Queensland Audit Department and later had a position in Works, Railways and Mines. The three years before he was sent to the Hodgkinson were spent in Charters Towers as the local Clerk of Petty Sessions and the Mining Registrar. Though the Hodgkinson men would have preferred a warden with the experience of Howard St. George they were pleased, initially at least, that their wardenless days were at an end.

For Jack, his troubles were only starting. A newspaper editor published that he had abused his trust and seduced the daughter of a friend. It alleged he was guilty of a string of disgraceful and corrupt actions and was of 'dissolute habits.' This was a heavy blow to the man described as the 'best man out of ten thousand'[53] on the Gympie field and who was little short of idolised on the Palmer. Jack protested his innocence and took the editor to court. Northern newspapers of that time considered libel costs to be quite normal running expenses and the editor was not greatly discomforted when the judge awarded Jack one hundred and fifty pounds damages. The Sydney *Bulletin*, in 1880, upheld Dr. Jack's good reputation and acknowledged the court judgment against the *Hodgkinson Mining News* and the awarding of the financial compensation but, it noted, 'the awarding and the obtaining of a verdict are two different things. The paper is printed too far north.'

The Hodgkinson was not the happiest of fields for Dr. Jack. He and Warden Mowbray were not bosom pals. One historian wrote that Jack and the warden 'quarrelled interminably'. By far, Jack had the more practical mining experience and after his stint at helping to draw up Gympie's famous Mining Laws and serving on the Miners' Court, he may have thought some of the warden's decisions ill-advised. In Parliamentary Votes and Proceedings of September 1879 a Select Committee dealt with the conduct of W.M.Mowbray in connection with charges against John Hamilton. The Committee decided that the late Colonial Secretary's correspondence vindicated Hamilton and that 'the conduct of the Police Magistrate at Thornborough (Mowbray) should be inquired into by the Colonial Secretary'.

53. *The Bulletin*, Sydney, April 4 1880

The Hodgkinson was also a reefing field as opposed to an alluvial one which meant capital investment rather than personal effort prevailed and Dr. Jack was ever the supporter of the little man, the underdog. However, miners are often a hard lot to please. Howard St. George, alias The Saint, was even unpopular on the Hodgkinson for one short stage.

SEEKING A ROAD FROM THE HODGKINSON TO TRINITY BAY

The miners agitated for a decent road to the coast that they could use instead of having to go back over the mighty Mitchell and its not inconsiderable tributaries to get to Cooktown. No one could tell the distance with accuracy but the popular estimate was sixty miles of reasonable road from Cooktown to Byerstown and then another seventy-five or so miles of rudimentary road over several wide river crossings to the Hodgkinson. Fortunately, telegraph lines to communicate with the Home Country were being built in the far north. One line already crossed the Seaview Range behind Cardwell on its way to Normanton and Cooktown was eventually connected by way of the Tate, Walsh, Palmerville and Maytown. A branch went from the Tate to the Hodgkinson in late 1877 and this line continued on to Cairns. Thornborough had a Post Office with James Mulligan as Postmaster a year before the telegraph arrived. The building housing the Post Office was Mulligan and Dowdell's store. A mail of a sort came from Byerstown or Palmerville but what miners wanted most of all was a decent wagon road and a port closer to hand than Cooktown.

A Cardwell butcher, George Clarke, for a subscribed one hundred and sixty pounds fee, found a track from the coast which followed the telegraph line and the Johnstone and Herbert watersheds. It then dropped into the headwaters of the Walsh River and continued to Thornborough. It was completed by mid-1876 and was said to be suitable for drays. The distance was approximately 140 miles and at least one gold escort used it to take gold to Cardwell and to bring back notes and coin in return. Clarke's memory lives on with a road, Clarke's Track, in the Malanda area. However, the distance seemed too great and Trinity Bay, the site of present-day Cairns was promoted as a port. Mowbray actively joined the search for a suitable track – and was condemned for leaving his warden's post to do it.

A meeting was convened at Thornborough with Howard St. George in the 'chair' (reportedly a pile of stacked logs) and a committee was elected to promote the road.[54] Port Douglas, Mourilyan and Trinity Bay were all eager contenders for the honour of the port site. John Doyle who had explored the ranges extensively

54. Dorothy Jones, *Trinity Phoenix*, Cairns Post, 1976 p58

with his party and who had seen the sea from a convenient ridge during his wanderings was asked to address the meeting. Mulligan, John Byers, Dr. Jack Hamilton, Williams and K.Murison were appointed to the committee charged with finding a party to locate a road suitable for wheeled traffic from the Hodgkinson to Trinity Bay. A subscription was started, then and there, and it was hoped the Government would augment the fund raised.

There is no doubt that John Doyle played an extremely important part in the discovery of this road but so many others participated that the real pathfinder is hard to name with certainty. Bill Smith, later of Smithfield, was one of the main players as was his mate Harry Evans, Inspector Douglas who was at that time in charge of the Thornborough Native Police and Sub-Inspector Johnstone from Cardwell. Others who played an active role in the search were a Warner who may have been surveyor Frederick Warner of Hann's party, a group of Fraser Island natives and the Hodgkinson Warden, Mowbray.

To Doyle goes the credit of discovering the Barron River. Mulligan had come across it but, because of the exceedingly close proximity of their headwaters, thought it was the head of the Mitchell. Doyle called his river the Silvers[55] and, as there was no watercourse marked on existing maps of the terrain between the Mossman and the Mulgrave Rivers, was ridiculed. The general opinion of the pundits was that, in the remote possibility of there being a river in that area, it would have to run uphill. Jim Mulligan was also derided. The Barron was often referred to as 'Mulligan's Mistake'. Geologist Robert Logan Jack stood by him, commenting it was a very easy mistake to make. The heads of the two rivers, though they enter the sea on different coasts are very close together. Some of the teamsters asserted that, on the divide, water in one wheel track flowed into the Barron while the storm rain in the other entered the Mitchell. The position of the Barron in regards to the Mitchell is an unusual one. It was skilfully utilised in the planning of the Tinaroo Dam which stores water from the eastern fall of the Dividing Range and delivers it without the use of pumps, to the western side of the range, for irrigation purposes.

Two things were sure. There were eight miles of scrub between the range and the sea that were almost impenetrable and to go up the range even crows would 'need breeching'. Robson marked a track following the Little Mulgrave down the range - close to the Gillies highway – but it was considered unsuitable for wheeled traffic due to its steepness.

55. Dorothy Jones, *Trinity Phoenix,* Cairns Post, p61

Cooktown and Port Douglas lost out in the race to become the new port and land speculators lost no time in hastening to Trinity Bay. The coast and hinterland buzzed with activity. Eventually, with the help of ten thousand pounds from the Government, a dray road was established and carriers Cummings, Stewart and Cochrane initiated use of the road. McNut loaded up both horse and bullock wagons at Walsh's Smithfield store bound for the goldfields but Maurice Fox, after being one of the first three teamsters over the track, decided not to try it again. Gold escorts used the road and in 1877 all but two thousand pounds of Cairns' 134,000 pounds worth of product exported was Hodgkinson gold. While on gold escort duties, Constable O'Dwyer was accidentally shot dead[56] and a sandstone headstone marks his resting place overlooking the Barron River.

The Hodgkinson Miners Back Dr Jack as Minefields Medico

Back at Thornborough, although Dr. Jack made no secret of not being a registered doctor and his patients were more than willing to let him treat them, others made trouble. As well as being the 'hospital' doctor, Jack also operated a private practice. Someone, calling himself 'A Miner', in a letter to the editor of the *Hodgkinson Mining News* of 2nd February 1878, complained about Dr. Jack holding the position of hospital doctor.

'From what I have heard I believe Mr. Hamilton is generally credited with being a man of abstemious habits, and no doubt diligent in the discharge of his duties to the best of his knowledge, provided always such duties are accompanied by remuneration; but the fact of him being sober, and even most zealous in the discharge of whatever duties may be committed to his care, is not sufficient, unless he possess the necessary legal qualification. His sobriety would not obtain for him the position or title of Captain in an Arctic Expedition, or Barrister in a Court of Law, and yet he is as much the rightful owner of either of the before-mentioned titles as he is to the assumed Medico one which he enjoys.'

A Miner went on to write that appointing Mr. Hamilton to the Hospital was a grave mistake and a good pretext for not subscribing until 'a duly qualified man is appointed.'

The question of Jack's status came to a court case heard by two Justices of the Peace before there was a 'duly qualified' magistrate available. This was on 23rd January 1878 and the case J.A.Hearney v J.Hamilton. Perhaps Mr. Hearney may have been 'A Miner'. The prosecutor had to prove Dr. Jack was not 'duly qualified'. Jack did not have to prove he was. The majority of the townspeople

56. Dorothy Jones, *Trinity Phoenix,* Cairns Post, p98

didn't want to lose the services of a doctor they'd long put their trust in and the proceedings were dismissed when the prosecution failed to make a case. They couldn't have heard Jack's deposition to Justice Sheppard. The two J.Ps were old Palmer men, Jim Mulligan and John Duff. Duff, Edwards and Leslie were the Maytown butchers and later established cattle stations in the area. Quite likely one or both of the justices was at one time treated by their Dr. Jack. For a while the infamous hospital was doctorless. Like hundreds of his patients Jack, himself, was stricken with fever and left the Hodgkinson for Cooktown to recuperate.

Mowbray had other problems besides Dr. Jack and the finding of a road to the coast. A young man, Alexander Dorsey, was assistant warden under Sellheim from 1874 to 1876 until a letter from Mowbray alleging Dorsey's misconduct forced him to resign. In August 1878, some fifteen months after leaving the goldfields, Dorsey encountered Mowbray in Queen Street, Brisbane. It wasn't a happy meeting. Dorsey asked Mowbray why he'd told such arrant lies about him. Mowbray denied that what he'd reported was anything but the truth. Dorsey slapped Mowbray's face. Mowbray retaliated, breaking his walking stick over Dorsey's head. A 'rough and tumble' ensued in which the two men 'scruffed one another by the hair'.Dorsey was arrested for aggravated assault and, 'in the interests of protecting every superior officer in the Civil Service' the Govenment, through its attorney Ratcliffe Pring, Queensland's first barrister, pressed that Dorsey be given a prison sentence.

The Magistrate, Mr. Pinnock, didn't agree. He thought the maximum fine for aggravated assault would serve and fined Dorsey five pounds – the same as Dalrymple's associate got for assaulting the Justice of the Peace in Rockhampton. According to folklore of the time, it led to a saying when someone felt a strong desire to thump his opponent, 'I've a good mind to take a fiver's worth out of his hide.'

The altercation doesn't seem to have irrevocably held back Dorsey's career in the Public Service. In May the following year he held the position of Clerk of Petty Sessions at both Clermont and Copperfield and was Registrar and Land Commissioner of Clermont.

THE COEN GOLDFIELD

There was activity again in Cooktown in 1878 when Robert Sefton, Sam Verge and party returned from Coen with 140 ounces of gold. Experienced prospectors like Mulligan, they were diffident about proclaiming a gold strike. Coen was even more isolated than the Hodgkinson and the quality of gold could not compare with the almost pure Palmer gold. They did not want to be responsible for starting a rush in such a remote area but, of course, the news got out.

A meeting was called in Cooktown and businessman Callaghan Walsh proposed that two hundred pounds be subscribed and given to the prospecting party to blaze a track to the field. Reluctantly, the prospectors agreed and the rush was on again. At first most diggers left on foot. Later others went by cutter to Port Stewart and overlanded from there but within six months most had returned. Coen wasn't altogether a write-off. It supported a small reefing field in which the quality of gold improved markedly. Over three hundred men were still employed there in the early nineties. It also led to the opening-up of the Hamilton field at Ebagoola by John Dickie and some minor fields in the furthermost north. Jack's grant helped considerably in funding the exploration and prospecting of these remote fields.

Dr Jack Hamilton, MLA for Gympie

By the late seventies 'Jack the Hatter' Macrossan was in Parliament. Hamilton was sick of the bickering on the Hodgkinson and when the Gympie miners contacted him to return and contest the forthcoming election as their representative, he decided to follow Macrossan's lead and try to help his miner mates from within the corridors of power. He defeated the sitting member, James Kidgell of Gympie's Two Mile, at the elections at the end of 1878 and in January 1879 took his place in the Eighth Parliament as John Hamilton, M.L.A., Member for Gympie.

The Eighth Parliament opened on January 14th with a tragic bang. As was the tradition, a salute was fired from muzzle-loaders in the battery at Queen's Park when the Governor left his residence to open Parliament.[57] After each volley, the barrels were cleaned ready for the next salvo but unfortunately a smouldering remnant lurked in one gun. As the second charge was inserted, the residium ignited the powder mixture and exploded prematurely, killing gunners Wilkie and Walsh as, at the cannon's mouth, they rammed home the charge. A grim start to the Eighth Parliament.

Dr. Jack was pleased to find his friend Macrossan, Member for Townsville, as Secretary for Public Works and Mines in the McIlwraith cabinet. Like Jack, Macrossan was a supporter of the 'little man' and was well-experienced in mining, its rewards and its tribulations. A fierce supporter of the North he was also an ardent Federationist. Now Dr. Jack hoped, something would be done for his

57. C. A. Bernays, *Queensland Politics During Sixty Years 1859-1919*, Govt. Printer Brisbane, p86

mates in the far north as well as for his constituents closer to the seat of power. He'd work indefatigably for Gympie interests but would support the Northerners in every way possible as well.

To his mates on the Hodgkinson and to his newly-elected colleague, Fred Cooper, Member for Cook and cousin to the Chief Justice, Pope Cooper, he'd already pledged to further the case for a decent road to the mining fields. Having been there and experienced the hardships caused by isolation, Dr. Jack knew what it was like when even the ubiquitous packers couldn't get through. Not only were miners and their families reduced to barely survival rations but necessary medicines were held up by flooded rivers and bog at the very time there was the greatest need for them. Fevers raged when the rain brought with it the mosquito hordes. No food drops or airlifted evacuations then.

When Jack entered Parliament it was then for a term of five years. The numbers in the lower house were increasing. Cook, a newly-created electorate, brought the number in the Legislative Assembly to fifty-five. None of these representatives received payment for their services though some travel expenses were allowable. This tended to keep the non-moneyed 'working class' out of Parliament. Knowing the vicissitudes of mining, Dr. Jack hoped the Queen of the North would continue to prosper. Unless a candidate could support himself during his parliamentary term there wasn't much point in his nominating.

WILLIAM REA, MLA FOR ROCKHAMPTON

One who tried but failed was quirky William Rea from Rockhampton, perhaps Samuel Griffith's most impassioned admirer. Rea was a gifted versifier with a light-hearted wit and lampooned his Parliamentary colleagues in verse. McIlwraith, his idol's rival, was a prime target and in Rea's maiden speech, an address in reply, in May 1879 he had a dig at the Premier's pro-kanaka policy in a jingle Charles Bernays preserved in his reminiscences of Queensland politics.

'Sing a song of sixpence
Our speech is all my eye.
It was only meant to humbug
And with a form comply.
But the four and twenty blackbirds
From the islands still we'll bring,
For white labour, we protest,
Is not at all the thing.'

The four and twenty blackbirds, of course, referred to McIlwraith's support for the plantation owners and their cry for South Seas labour.

Rea was a well-liked, chirpy little fellow. His Parliamentary colleagues were shocked at his death in 1881 to find he'd been subsisting in abject poverty, with no money to spare for medications and treatment, his cheery nature completely cloaking his wretched distress.

PREMIER MCILWRAITH

Scottish-born Premier McIlwraith was a civil engineer. After a stint at store-keeping he joined the Victorian railways as a surveyor and rose to the rank of engineer. He left Government service to become, in time, partner in a firm of railway contractors where his engineering knowledge was largely responsible for their success in the difficult terrain. This grounding in railway construction made him aware of the potential of a railway network in a growing Colony like Queensland. To him, the transcontinental lines of Canada and the U.S.A. built on the land grant system were an inspiration. He could see clearly what a later Australian Prime Minister called The Big Picture. Whether the vision would be practical or not is an entirely different matter.

With his Victorian profits he bought pastoral runs in Queensland and the Northern Territory and was a major shareholder in a pastoral company that survives today despite droughts and myriad reverses – North Australian Pastoral Company. Besides railways, he was vitally interested in pastoral matters and a keen supporter of the idea of vertical integration, expanding his interests to encompass freezing works to take advantage of the new export opportunities opened up by T.S.Mort's experiments in the sea transport of frozen meat.

A case which was tried in the Rockhampton court when grazier Peter MacDonald claimed a Queensland Government irregularity in the issue of grazing runs, led to a verdict for the plaintiff.[58] He successfully claimed that, after the runs had been rented to him, the Government then re-leased them to other occupiers. He was awarded a staggering amount, almost twenty thousand pounds which included a sum of seven thousand pounds interest on the money. Fortunately for McIlwraith's government the misdemeanour occurred before they came to power.

From Cairns, the irrepressible Archie Meston was earning himself the soubriquet 'The Sacred Ibis' because of his penchant for quoting from Egyptian and Greek mythology. This was to the despair of the Hansard clerks who were not at all well-versed in mythology and found the spelling challenging at best. Fred Cooper and Lumley Hill who also came into the Eighth Parliament figured with

58. C. A. Bernays, *Queensland Politics During Sixty Years 1859-1919,* Govt. Printer Brisbane, p76

Jack and Tom Campbell in a later voting controversy in the Cook electorate. Jack's Gympie colleague Horace (now Sir Horace) Tozer was Queensland's Agent-General in London.

McIlwraith didn't come into the Eighth Parliament as its leader.[59] That honour went to John Douglas, a mild-mannered man who was descended from the Marquess of Queensbury but who lacked that gentleman's knowledge of fighting – in this case, political in-fighting. Douglas spurred the Northerners' hopes in July 1877 when the Colonial Treasurer, Dickson, introduced a Bill to 'divide the Colony into Districts for financial purposes, and to adjust the general and local receipts and expenditure of the Colony.'

This wasn't exactly pro-Separation but was intended to give the new districts financial autonomy. It passed the second reading but disappeared into oblivion leaving the Northerners to bemoan the fact that the wealth obtained from their goldfields followed the laws of gravity too well and gravitated down to fill Brisbane's coffers. Very little made its way back up north. In May 1879 during McIlwraith's Parliament, Macrossan introduced a similar Bill but it met with an even more rapid demise. In September the order for a second reading was simply dropped from the parliamentary agenda.

Douglas lost his leader's position early in 1879 after the Address in Reply. Thomas McIlwraith as Member for Mulgrave challenged for the leadership and won in a re-sounding victory. He held the Colony's top job, on and off, until he, too, was ousted over an unpopular decision in 1893.

ERNEST FAVENC'S EXPEDITION FROM BLACKALL TO PALMERSTON (DARWIN)

The beginning of McIlwraith's first year as Premier saw the return to Brisbane in the S.S. *Ocean* of Ernest Favenc and his co-explorer Briggs, of the Queenslander Expedition from their successful foray from Blackall to Palmerston (later Darwin). They were to find a route for the proposed railway to link Queensland with the far northern port – a Transcontinental Line in the grand tradition of rail. The third member and second-in-command, G.R.Hedley followed later after recuperating at Powell's Creek Station. He'd finally succumbed to the fevers that had plagued them all and was too ill to travel with his companions when they left.

For a short period the party had a fourth member, Opal, an Aborigine and ex-Native Policeman who joined them at Cork Station and disappeared not long after they met up with a large party of Aborigines at Corella Lagoons. These were demonstrably friendly, the lagoon offered a tantalising menu of bush delicacies

59. C. A. Bernays, *Queensland Politics During Sixty Years 1859-1919,* Govt. Printer Brisbane, p85

and the explorers decided Opal had most likely opted for tribal company rather than the austere life with the explorers. The expedition was sponsored by the magazine, *Queenslander*, whose editor Gresley Lukin, a friend of Favenc's, thought a small select party would have more chance of success than a cumbersome expedition. He was right.

They were by no means the first explorers to venture into that area. It had already been traversed by the Burke and Wills expedition, Leichardt, Landsborough, the Gregorys and others. Pastoralists like the Duracks, Buchanan and Frank Scarr provided useful information about the travellers' proposed route. The Overland Telegraph Line was almost completed, giving the explorers a means of communication providing, of course, they lived to make contact with the Line. It took a month to get from Blackall to Cork Station on the Diamantina and by the time they did reach the Line and were able to get a message through, their Brisbane supporters had all but given them up for dead.

Favenc's diary was not the serious kind expected of explorers but he did try to uphold the standards.[60] 'I like sticking in readings of thermometer,' he wrote. 'It looks scientific; and I have noticed in all the late journals of exploratory expeditions, that they do it.' He also described for laymen the difficulties of traversing the spiny spinifex by explaining, 'imagine about 5000 knitting needles rolled up in inextricable confusion, with all their points outward.'

Nevertheless, their expedition was not a picnic outing. They encountered extreme heat and suffered thirst repeatedly. Horses, including Favenc's favourite, were lost from heat exhaustion as well as from lack of feed and water. The explorers lived for much of the time off the land. Hedley was the chief marksman and their diet included parrots, ducks, pigeons, one emu, one native companion and fish. 'Kangaroos and other marsupials' were noticeably absent. The scarcity of kangaroos was also noted by Gregory in his diary of his trip to Queensland from Port Essington. Favenc continues, 'Crows and hawks were carefully reserved to the last when all else should fail' but they did, at times, dine by necessity on eaglehawk and frogs.

On Christmas day, with rations low, they had a pot of tea and two tablespoonfuls of flour made into a 'skilly' for Christmas pudding. 'Then filled ourselves with pigweed.'

Water was in very light supply and birds which could have provided the traditional festive poultry had fled but – 'Heavy rain at sundown. Two frogs

60. Cheryl Frost, *The Last Explorer*, J. C. U., 1983 p14

appearing, I killed and toasted them,' Favence wrote, 'Although very small they are awfully good. Turned in.'[61] Christmas was over for another year.

Following the rain, the weakened horses had problems with bog but in the second week of January they struck the Overland Telegraph Line and followed it south to Powell's Creek Station. It had taken them much longer than they had anticipated so the telegrams they were able to send were very well-received by anxious relatives. The Wet also brought with it recurrent fevers and the men were increasingly grateful they were now in 'civilised' country.

THE DROWNED MINES AT THE PALMER AND THE CYCLONE OF 1878

As well as Favenc, another of Dr. Jack's northern friends was in town. W.H. Corfield, having had a particularly bad attack of malaria, had taken time off from the goldfields to try a cure in the bright lights of Sydney. The cure presumably worked and he was in Brisbane on his way back to the north. What he told Jack of the mines on the Normanby, the eastern part of the Palmer field, caused some concern. Water was coming in and the pumps available were unable to lower the water level. The overseas mine-owners ordered the pumps pulled out and the mines abandoned.

'What of the Palmer?' Jack asked, anxious for the prosperity of the Queen of the North that footed the bill for him in Parliament.

'Not yet,' replied Corfield in an enigmatic tone and changed the subject.

Word had come while Corfield was in Sydney of the death of another adopted Northerner. Richard Daintree died in June aged forty-six. In the same week, though ten thousand miles away, Daintree's mentor, the father of Australian geology, the Rev. W.B.Clarke, also passed on. He was twice Daintree's age.

Jack worried for nights over the possibility of water stopping his mining operations. Drowned mines were something he'd already experienced in the early days of Gympie. He soon had something else to cause him concern. 1878 saw the infant Cairns almost demolished by a severe cyclone. The winds rushed in from the Coral Sea early on the 8th March. The flimsy structures offered no opposition to their force. It was the end of Smithfield which had been a virtual ghost town since many of its inhabitants deserted for Cairns after a crippling flood the previous year. Sheets of iron from the shanties along the shores of Trinity Bay were lifted, swirled high and unceremoniously dropped out at sea. The unrelenting rain caused landslides which devastated the hillsides and the

61. Cheryl Frost, *The Last Explorer*, J. C. U., 1983 p14 p23

accompanying noise and fury as the ground shifted, made the beleaguered citizens fear an earthquake had joined forces with the cyclone.

For the ships in the harbour there was no Met. Bureau warning. Most were alerted too late by their own falling barometers. Some sought refuge in the open sea. Others took their chances close to shore. Many were lost. The *Kate Conley*, with a load of Daintree cedar sailed out and vanished with a crew of eight. Two years later, Logan Jack found her battered hull in Temple Bay, closer by far to Thursday Island than to Trinity Bay.

Queensland Claims the Coastal Islands North to New Guinea

McIlwraith, ever wary of our neighbours to the north, brought in a Coastal Islands Bill, claiming for Queensland all the islands within the Great Barrier Reef and the Gulf of Carpentaria plus those in the Torres Straits north to the continent of New Guinea. When Queensland gained Separation from New South Wales, her rights to the coastal islands were not clearly stated – 'all and every adjacent islands'. The question in McIlwraith's mind was 'How adjacent is adjacent?' In 1872 the limit was set at within sixty miles of the coastline. The settlement at Port Albany, named Somerset after the First Lord of the Admiralty, was established after Governor Bowen toured the northern ports on a voyage of inspection. On his return, he recommended in a despatch to the Imperial Government that a settlement be made at Port Albany on account of its geographical significance. It was to be a 'harbour of refuge', a coaling station and a centre for trade with the lands to the north and to the islands in the North Pacific. A detachment of marines was sent out to the site and Jardine, the Police Magistrate in Rockhampton (whom we have met previously) was sent to supervise the establishment of the outpost.

It was Jardine's assessment that an outpost so isolated would need to be reasonably self-sufficient and he arranged for his two sons, Frank and Alick, aged twenty-two and twenty, to overland a mob of cattle to form a station from which meat supplies for the settlement could be easily obtained. The Government thought it a good idea and agreed to supply the party with a qualified surveyor and his necessary equipment.

The brothers finally made it, but with great difficulty and after losing almost all their horses – a great many from the effects of a poisonous bush – and a fifth of their cattle. They also lost most of their supplies and belongings when a carelessly prepared fire burnt the camp out in their absence. No human lives were lost and though the trip was a financial disaster for the Jardines who refused to accept compensation for their losses from the Government, it was undoubtably a success story for the two young men.

Unfortunately the choice of Somerset as a port was a mistaken one. The Torres Straits were excessively dangerous to shipping with submerged reefs, treacherous currents and rocky islets. Wrecks that were found showed that ships had come to grief there even before the Dutch sailed through. The pearling vesels that followed the settlement chose Thursday Island as their headquarters and with no hinterland to support it, Jardine's Somerset fell behind in the race to become the centre of commerce, an antipodean Singapore.

A little further to the south, in the Cairns hinterland, discoveries were being made that led to the continued growth of that centre. Atherton found tin at Tinaroo and a copperlode was located near Herberton. Robert Logan Jack left Cooktown seeking – and finding – agricultural land to the north. All this, together with the news that J.Thornloe Smith had already prepared plans for the start of the Cooktown to Maytown railway as far as the Normanby siding, put Dr. Jack in a more optimistic frame of mind.

6

SCANDALS & SQUABBLES. THE NINTH QUEENSLAND PARLIAMENT DEVELOPS THE FAR NORTH 1883

Special taxes on Chinese miners on the Queensland goldfields. Direct mail steamer route from London to Brisbane. The Transcontinental (Land Grant) Railway Bill passed. Hemmant Iron Rails Inquiry. McIlwraith and Griffith feuding. The tragedy of Mary Watson and baby and Ah Sam near Lizard Island. Gold at Mt. Morgan. First mailcoach to Herberton. New Guinea annexed. Gold at Coen. £10,000,000 Loan Act. Philp in parliament. Halpins/California Gully poll. Work commenced on the Cooktown/Maytown rail. Hodgkinson and Palmerston's defence. Griffith moves for Federation and brings in payment for M.Ps.

QUEENSLAND GOLD EXPORTS AND THE GOLDFIELDS ACT

Big things were happening in Australia and elsewhere in the world in 1883. Edison invented the electric light bulb. Melbourne installed Australia's first telegraph exchange. In Sydney *The Bulletin* began its long and fruitful life. Gold was discovered on the Margaret River in the Northern Territory and miners rushed once more. Jack and Newell found tin at their Great Northern Mine on the Wild River at Herberton. Sugarcane was grown at Innisfail and the railway line from Brisbane reached the western township of Roma. In Brisbane, the people's representatives were hard at it trying to run the Colony.

Seasons hadn't been kind to the pastoralists but gold exports made up to the colony the income that was lost from wool and meat. Periodically, the Goldfields Act was amended with both Macrossan and Dr. Jack putting forward practical suggestions. Queensland repealed the Act that required the Collector of Customs to collect duty on gold found in the Colony, almost as soon as it gained Separation. This cut off what could have been a huge revenue-raiser. Looking back to his early Gympie years, Dr. Jack was proud to have been one of the committeemen who prepared the sixty-eight pages of Rules of the Gympie Local Mining Court. It gave him invaluable experience in understanding as well as formulating mining laws and remained the standard mining guide for many years.

Quite a few amendments to the Goldfields Act passed a first reading but fizzled out like a duffer claim by the second. When George Thorn's Ministry held power, another effort was made to stem the influx of Chinese miners – mainly to the Palmer where there were an estimated 7,000 of them in the mid-1870s. A Miner's Right cost ten shillings but to an Asian, the same Right had to be purchased at six times that amount. Similarly registrations for business licences discriminated against the Asians and Africans who were legally required to pay double the fee payable by a European. Unless a Chinese or an African actually discovered the goldfield, miners of both races were excluded from the field for the first two years of its life.

Though the amendment to make life even more difficult for these two races passed all stages in Parliament, the Governor withheld consent. New South Wales and Victoria had imposed Poll taxes on Chinese migrants but Governor Cairns felt it would be unfair to the growing number of Chinese/British subjects in Hongkong, Labuan and the Straits Settlements. Griffith, a more than competent lawyer, was at this time Attorney General and put his view to the Governor that, as the Colony's democratically elected Government had approved the legislation, the Governor should ratify it. Cairns disagreed with his argument.[62] Once again, the Governor over-ruled the elected Government.

Queensland rallied support from the other colonies. New Zealand was also approached but diplomatically declined to express an opinion on the rights of self-government in a sister colony. It was a noble though futile stand against imperialistic power but it led only to a minor political upheaval and a subsequent change of Ministry. Griffith resigned.[63]

62. Eds. Murphy and Joyce, *Queensland Political Portraits,* U. Q. P. Brisbane, 1978 p149
63. Eds. Murphy and Joyce, Op Cit, p149

Another amendment to the Mining Act went through undefeated. This extended the life of a 'new' goldfield to three years after the date of its gazettal and no Asian or African could be employed on the 'new' new field until that time had expired.

Still not satisfied, first Macrossan and then Dr.Jack, both experienced miners, tried to bring in more amendments to update the Act, not necessarily trying to restrict non-European miners. None got past the crucial second reading until 1881 when Macrossan succeeded with a Mines Regulation Act, a complicated and detailed affair dealing with mines employing more than six workers underground. No boy under the age of fourteen nor any female was to be employed underground. Elaborate provisions were made for safety inspections and checks brought in to confirm the payment of correct wages. The Act benefited collieries as well as underground gold mines and had a comparably smooth passage through the Assembly, capably and diplomatically steered by Macrossan.Gold production continued strongly and for this, with other industries not doing so well, the Government was again duly grateful.Legal gold exports rose from a healthy 49,091 ounces in 1867 to just under 400,000 ounces, though they did fall back to 260,000 ounces in 1881.

The Queensland Railways Scandal of 1879

Back in 1860 Premier Herbert was lamenting that the 'English mail steamers no longer proceed beyond Melbourne' and that Brisbane needed a more direct route and a speedier mail contact with Britain. He suggested a route from the north-west, past the East Indies, through the Torres Straits and down to the capital but nothing came of it until Thomas McIlwraith commanded the action in the reign of the Eighth Parliament. Herbert's idea fitted in perfectly with his Big Picture plan for developing Queensland from the Top down. In his characteristic masterful manner he overrode Griffith's objections and, although the Government through fifteen sittings had failed to ratify the contract for the mail, neither had they cancelled it. McIlwraith decided to take the initiative and cabled the company involved to 'proceed with your arrangements' to carry out the service. In no time at all, Brisbane had its own direct mail link with London.

The highlight of the year was the Premier's Transcontinental (Land Grant) Railway Bill.[64] It authorised the construction of a line from the vicinity of Charleville to the Gulf of Carpentaria with branches to Hughenden and Cloncurry. Seven and a half years were allowed for the construction. Unhappily it

64. C. A. Bernays, *Queensland Politics During Sixty Years 1859-1919*, Govt. Printer Brisbane, p91

never came to fruition and led to a bitter and long-lasting feud between Griffith and McIlwraith, both exceptional men.

Tenders were called for the provision of steel rails, bolts, dog spikes etc. and for their freight to Brisbane, Rockhampton and Townsville.[65] In late 1879 a memorandum of agreement was drawn up between Ibbotson Brothers and the Minister for Works, Macrossan, for 42,000 tons of rail. At that time, the price of steel rail was down and hence it was an ideal time to commence extending the rail network. The Premier went to London to arrange finance, stopping off in America again on a 'fact-finding tour' to inspect their transcontinental rail link. While he was away a further tender was received from another London company, McIlwraith, McEachern and Co., in which the Premier had an interest. However, their tender was higher than that provided by Ibbotson's Brisbane agent, Thomassen, and Macrossan rejected it. A local Brisbane company considered tendering but were dissuaded by the low price quoted by Ibbotsen Bros. through their agent Thomassen.

Thomassen then entered into a contingent agreement with McIlwraith and McEachern to supply the Government with 30,000 tons of rail at six pounds per ton. In the meantime, McEachern in Brisbane, informed Andrew McIlwraith and his fellow principals in Britain that he had entered into no agreement with the Queensland Government for the rails but only one with Thomassen (Ibbotson) for their freight to Queensland.[66] The partners owned a shipping line which was bringing migrants to the colony of Queensland at that time.

Thereupon, Thomassen disclaimed the existence of all deals and pleaded his innocence by alleging that 'defective' telegrams conveying the wrong message were responsible for the confusion. On the strength of this, and with a rising market, Andrew McIlwraith decided he'd better try to sell the unwanted rail. He did, selling 25,000 tons which the company had bought in at six pounds a ton for over nine pounds per ton thus making a handsome profit from Thomassen's misconception. It was at this point that Premier McIlwraith arrived in London, a few days before Christmas 1879.

The price of rails was still rising and he was understandably upset by the confusion between the Ibbotsons and their Brisbane agent over the rail purchase. They refused to ratify the agreement which he considered they had made, and made willingly, in Brisbane. To top it all, Andrew had now re-sold most of the rail. It was of secondary importance that in doing this he'd made a hefty profit. Acting on his own responsibility and taking into account the rising market, McIlwraith

65. C. A. Bernays, *Queensland Politics During Sixty Years 1859-1919*, Govt. Printer Brisbane, p87
66. Royal Commission 1881, *Mr. Hemmant's Petition*, Govt. Printer

decided to 'purchase only such a quantity of rails as would supply the present wants of the Government.'

The Agent General was told to call new tenders for 15,000 tons. Ibbotson Brothers and seven other companies tendered. This included all the companies who usually contracted for the Colonial Government. Haslam Engineering Co. secured the contract, their price being nine pounds eighteen and six, one pound seven and six below the next lowest tenderer. Not being a manufacturer of rail and not having previously tendered to the Queensland Government, they purchased the rails from Barrow Haematite Co. and the Moss Bay Co.

Unfortunately these were the two companies Andrew McIlwraith had contracted for his rails and this led to allegations that they were the same rails McIlwraith and McEachern had unloaded earlier at the substantial profit.

Back again in Queensland the Government split into pro-McIlwraith and pro-Griffith factions. William Hemmant, on the Griffith side, petitioned for a Royal Commission into the deal and a Select Committee of Parliament also probed the matter. Both decided against the allegations of Hemmant, Dickson, McLean and Griffiths but some mud stuck and it didn't do McIlwraith or his proposed grand rail network any good. Even Willy Rea wasn't happy with the outcome. A man without prejudices, he considered Queensland itself was the biggest loser in the deal. Macrossan and Jack Hamilton stuck with McIlwraith. McIlwraith and Griffith feuded on. Bernays considered them both eminent men, outstanding in their time, but he made the distinction that while one achieved success 'with an oilcan', the other used a 'firestick' approach.

As could be expected, the air in the House was emotionally supercharged. Some of the members lost all control. The Chamber could not be contained. It sprung a leak. John Douglas decided to confide his point of view to the press.

The Speaker brought the House's attention to a letter in that day's *Courier* which 'purported to furnish a précis of certain Evidence which had been given before the Select Committee of the House, now sitting to enquire into the allegations contained in Mr. Hemmant's Petition.' The letter bore the signature of the Honourable Member for Maryborough, Mr. Douglas,.

Douglas admitted that he was the author.

McIlwraith moved that the Honourable John Douglas be found guilty of contempt.

There were the usual shenanigans and Douglas, the following day, remained unrepentant and declined 'to apologise for the contempt of which he had been adjudged guilty.'

McIlwraith, his Scottish dander up, firestick at the ready, moved that 'the Honourable member for Maryborough, Mr. Douglas, be committed to the custody of the Sergeant-at-Arms for removal from the House'. Griffith, his oilcan firmly in his hand, stilled the troubled waters so that cooler heads prevailed, the motion was withdrawn and the Honourable Member for Maryborough was not ignominiously apprehended but remained seated in his customary seat.

When Parliament was restored to a more even keel, they co-operated to such a degree that the Torres Straits Marine Pilot Service was instituted and brought into use. This service proved to be a boon for ships travelling the dangerous waters inside the Great Barrier Reef for many decades.

In the far north things were looking up although Cairns had been temporarily demoralised by cyclone and floodwaters. The discovery of tin at Herberton almost guaranteed Cairns a rich and profitable hinterland to support the growing port. Heavy and sustained wet seasons played havoc with the unsealed roads, often leaving the residents of the outlying settlements precariously short of the necessities for survival. A more reliable means of communication in the form of a railway was mooted but response from Brisbane where railways weren't the most popular topic for discussion, was luke warm.

THE TRAGEDY OF AH SAM AND MARY WATSON

Cooktown was in a state of shock when the captain of the *Kate Kearney* reported finding skeletons of two adults and an infant on one of the Howick Group islands.[67] They were easily identified. The taped pigtail was once part of Ah Sam who was employed by Captain Robert Watson in his bêche de mer enterprise on Lizard Island. The female skeleton, its bony fingers clutching the fragile fingerbones of her dead son, was Watson's young wife, Mary. The child, born six months earlier in Cooktown was their baby, Ferrier.

Captain Watson and his partner Fuller had decided to transfer their operation from Lizard Island to Night Island about two hundred miles further to the north. Bêche de mer were becoming harder to find at Lizard and Night Island promised a richer harvest. It was a logical decision and Mary asked if she could remain on Lizard with Ferrier, then only twelve weeks old, until a home could be made for her at their new base. Is seemed a practical alternative. Robert left Ah Leong and Ah Sam to look after his wife and baby and the men hoped to be back at Lizard within six weeks. After that, they should all be in their new home on Night Island in time for Christmas.

It didn't work out that way. Ah Leong was fatally speared in the island's vegetable garden and Ah Sam terminally injured with seven spear wounds. The young wife removed the spears, sutured the wounds as best she could with a

67. Jillian Robertson, *Lizard Island,* Hutchinson, p162

curved needle from her workbox and bandaged them with strips of torn sheet. Fearing a further attack, she gathered what provisions she could find and launched herself, baby Ferrier, Ah Sam in the metal tank they used for boiling the bêche de mer into the uncertain waters between the Barrier Reef and the coast. Optimistically, she took her jewellery and what money she could find – two pound notes and some silver.

Passing sailors noticed fires on the island, pulled in and saw native canoes on the beach and natives using the Watsons' hut. As the Watsons were known to be moving to Night Island, they naturally assumed they had already moved on.

The fugitives left Lizard Island on Sunday afternoon, September 2nd 1881.[68] Mary kept a tragic diary of their last days. A vessel passed close enough to raise her hopes and she hoisted the baby's pink and white shawl as a distress flag but no one noticed. Slowly the boat glided past and disappeared over the horizon. They drifted onto an islet in the Howick Group but could find no water anywhere. The last diary entry of September 11th read, 'Ah Sam preparing to die. Have not seen him since 9. Ferrier more cheerful. Self not feeling at all well. Have not seen any boat of any description. No water. Near dead with thirst'.

Mary came to Cooktown from Brisbane to tutor M. Bouel's children and, although she wasn't led to expect this additional duty, to play the piano at her employer, French Charley's restaurant. The men from the *Kate Kearney* made their heartbreaking discovery on 16th January 1882, the day before what should have been her twenty-second birthday. Ironically, when the mate of the *Kate Kearney* first saw the tank, it was partially filled with rainwater.

QUEENSLAND'S ANNEXATION OF NEW GUINEA IN 1883

There was excitement when gold was found at Mt. Morgan, also in 1882. The mountain proved to be almost solid gold and though the ironstone put prospectors off the track for some time, the wait was well worth it. McIlwraith travelled to the Palmer River to hear what the miners had to say and to give his attention to the rail line proposal from Cooktown to Maytown, the Palmer's commercial centre. While on the Palmer the Premier learned of his newly awarded knighthood. This he had tactfully tried to avoid though he accepted it with grace. He had the Scottish belief that self-worth meant more than inherited superiority.

Dr. Jack also visited his old haunts on the Palmer. The output from the Queen was declining so he and his partners decided to sink another shaft to try a new area of the Mountain Maid Reef. It was five years since he'd taken out the lease and the

68. Jillian Robertson, *Lizard Island,* Hutchinson, p162

existing workings were declining in productivity. By mid-1882 he was in Herberton to welcome the arrival of the first mail coach using the route he, as a committeeman, had helped plan.

Before McIlwraith pressed on with his plans for the Transcontinental (Land Grant) Railway Bill, he paused for an equally bold venture in annexing New Guinea for Queensland. At the end of the previous year (1882) it was brought to his notice that an article in the *Allgemeine Zeitung* recommended that Germany annexe New Guinea as a colony without respect to the Dutch occupancy of the western half of the island. In February the Premier attempted to bring this matter to the notice of the Imperial Governor, Sir Arthur Kennedy, and cabled the Agent General in Britain to urge the government there to out-manoeuvre the Germans with an annexation of their own. Queensland, he promised, would bear the expense of the enterprise if only for security reasons.

He received no reply to his urgent petitions. Kennedy was in poor health and subsequently died in Aden a few weeks later on his way 'home'. The Imperial Government did not deign to reply. Acutely conscious of the risk of a foreign power taking over a near-neighbour and impatient at the diplomatic delays, McIlwraith acted alone. On 20th March 1883 he telegraphed the Police magistrate at Thursday Island, a Mr. Chester, to steam posthaste to New Guinea in the *Pearl*, there to take possession of that part of New Guinea not already controlled by the Dutch. Fourteen days after receiving the cable, Chester was able to send his reply – Mission Accomplished. Chester and the *Pearl* won the race.[69] The German corvette *Carola* left Sydney two days before the magistrate received his cable but distancewise, Chester had too much start on it. Although Queensland won the race, it wasn't awarded the prize. Months later, after carefully thinking the matter over, thc Imperial Government refused to ratify Queensland's 'impudent' act on the grounds that it would not be viewed as a 'friendly act' by other powers.

Naturally McIlwraith was furious at yet another snub by the Imperial Government and contacted the other colonial premiers. Not usually noted for their desire to co-operate, on this occasion all colonies sent their unanimous approval and, united, appealed to Lord Derby to reconsider. In late 1884 a British Protectorate was eventually proclaimed over the south-eastern section of New Guinea and the Australian colonies guaranteed to provide fifteen thousand pounds annually towards the costs of administration. Queensland readily agreed to fund, on a population basis, its share of this amount.

---

69. C. A. Bernays, *Queensland Politics During Sixty Years 1859-1919,* Govt. Printer Brisbane, p92

SURVEYOR EMBLEY AND THE OVERLAND TELEGRAPH LINE FOR FAR NORTH QUEENSLAND, 1881

The first Special Commissioner for the Protectorate, Sir Peter Scratchley, took up his position in 1885 but before he'd seen twelve months service died of malaria. His place was taken by a Queensland ex-Premier, John Douglas. He was the man whom McIlwraith had so successfully challenged for the leadership of the colony. With the annexation of New Guinea safely accomplished, McIlwraith again turned to his vision of the railroad to the Gulf. As Minister for Works in the Macalister government he had brought the matter up in Parliament. Macalister, by-passing the normal channels of Cabinet debate, leaked to the *Courier* that he was vehemently opposed to the idea. McIlwraith resigned from his portfolio as a result. In 1883, as Premier, he tried again. A group, the Australian Transcontinental Syndicate, was proposed under the leadership of an eminent army man, Major General Fielder, who, with a considerable entourage, came to inspect the probable route. He was impressed by the scheme and gained support for the syndicate from the Earl of Denbigh and a number of financiers, gentlemen who, regrettably, had Semitic-sounding names. The populists of Queensland and the squatters who resented losing as much as an acre of their land to the railway, were horrified and protested loudly. As Bernays reported, they thought it savoured of a Jewish invasion under the leadership of a 'distinguished earl' and an 'eminent soldier'.

Queensland was vigorously opposed to the American style of railroad financed by land grants along the route. The knockers argued rail lines should follow population growth and not attempt to initiate it. Much of the land to be traversed was unsuitable for anything except grazing but the setting-aside of reserves for future railway lines in what was described as prime agricultural land was not looked upon favourably either. It was a bold scheme especially in those times. Few besides McIlwraith were prepared to risk all in its cause.

Jack's thoughts were constantly involved with mining when he wasn't supporting his Premier in the House. He never lost a fervent hope that gold would be discovered in the farthest north and was cheered considerably when Sefton and his mates found gold in the Coen area. It wasn't in the same class as the Palmer or even the Hodgkinson but Dr. Jack had faith that some day, someone would strike the rich motherlode that gave birth to the scattered outcrops. Robert Logan Jack ardently believed gold would be found in the metamorphic rock under the top layer of sandstone in the far north and our Dr. Jack held the geologist Dr. Jack in the highest regard. New Guinea was mining profitable gold. Both Dr. Jacks could not see why the far extremity of the Peninsula could not be brought to follow suit.

Dr. Jack was optimistic when Surveyor John Embley and party left the Hann River in Princess Charlotte Bay where Donald Mackenzie had set up his Lakefield cattle station in 1881. Mackenzie was friendly with the natives living in the area and when necessary offered them protection from less humanitarian white men. However while routinely applying ointment to a horse's saddlesore, an Aborigine known to him came up to him casually, apparently to watch the procedure. Unexpectedly, without reason or warning, the Aborigine speared Mackenzie through the body. He recovered from that attack but was clubbed to death many years later when his men were out mustering and he was working alone in his garden.

Embley was assisting in the surveying of a route for the construction of the Cape York Telegraph Line that was a high priority with almost everyone in the colony.[70] It was one venture which gained approval from all factions regardless of their political leanings. Once the prospectors ventured into the northern part of the Peninsula, cattlemen followed. Glen and Charles Massey took up Lalla Rookh (now Silver Plains) near Port Stewart two years before Embley's expedition. Within three years Charles was fatally speared and the surviving brothers Glen and John formed an outstation further inland at Rokeby.

Embley and his party headed west commenting on their way on the different forms of Aboriginal architecture. The Princess Charlotte Bay tribes built 'well-constructed bee-hive gunyahs designed to protect them from the mosquitoes' which were a serious worry close to the coast. As they moved away from the coast the bee-hives gave way to a 'fragile platform, or often two platforms, of sticks forming a sort of two-storey sleeping place without walls. The lower platform is occupied by the woman whose duty it is to keep up the smoke-generating smouldering fire. The man reposes on the upper platform, reaping the benefit of the smoke but taking no part in the work'.

Along their progress up the western fall they encountered a tree marked J and other remains leading them to believe they were re-visiting the Jardine brothers' camp sites. Embley named the Edward River after his brother.[71] Further north another stream perpetuates his own name. Around Edward River they found the explanation for the platforms on uprooted tree stumps they had sometimes seen in swamps.They were used to protect corpses, well-secured in tea-tree bark wrapping from the ravages of dingoes and voracious ants.

There was some difficulty in recognising the identity of an occasional river named by the Jardines. The brothers and their surveyor, Richardson, often

70. R. L.Jack, *Northmost Australia Vol 2,* Robertson Melbourne, 1922 p631
71. R. L.Jack, Op Cit, p648

travelled at a distance from each other, the former taking the route more suitable for the cattle, the other a more direct path with the camping gear. Embley found it difficult at times to reconcile their diaries but he carried on. His was the definitive naming. Travelling more to the north-east he crossed tracks made by Masseys' dray as they journeyed from Lalla Rookh to Rokeby. Here the country was good, the travelling easier and the party followed Masseys' dray tracks up the Coen River, past the then-deserted Coen diggings. Embley in time, according to Robert Logan Jack 'traversed and surveyed most of the important creeks and rivers as well as the boundaries of reserves and pastoral leases within the region now under consideration.'

McIlwraith took the isolated nature of northernmost Queensland to heart, for as well as being that part of the colony furthest away from the seat of Government it was also the part closest to world markets. Somerset was settled, New Guinea annexed and even a railway line envisaged from Somerset to link up with lines in the settled areas. This proposal had universal support. South Australia already had its Overland Telegraph Line and McIlwraith was determined that Queensland would soon enjoy that form of international communication as well. John Bradford, Inspector of Lines and Mail Route Services, was to find a suitable route commencing at the Laura/Cooktown line at Cooktown and proceeding to Cook's landing place at Possession Island. Previous explorers had ruled the project feasible and Bradford with his party of five whites, Jimmy San Goon and Johnnie, an Aborigine, with their 36 horses (of which 13 lucky ones made it to Somerset) assembled in Cooktown. Each man had a Martini Henri carbine and a Colt revolver. Healy, the second in command, also carried a fowling piece. They left Cooktown June 6th 1883 and reached Somerset without major problems on August 29th, not long before a government-changing election. A 'satisfactory route, almost a straight line, was decided upon,' wrote Logan Jack, 'and construction commenced.'

THE 1883 ELECTIONS IN FAR NORTH QUEENSLAND

When the Ninth Parliament opened in November 1883, McIlwraith's arch-rival, Samuel Griffith, the very astute lawyer who did his articles at Ipswich under another Premier, Arthur Macalister, led the Parliament. At thirty-eight, he was more of an age with Jack Hamilton and was ten years McIlwraith's junior. A successful Brisbane barrister, as was the custom in those days, he continued his practice while he was in Parliament. After all, at that time, he was one of an elite company. There were only twenty-five barristers in the colony to cater for a population of about 300,000 and there was no Conflict of Interest clause to stop him.

It was in Griffith's first term that the Ten Million Pound Loan Act came before Parliament though actually the amount sought was a mere 9,980,000 pounds to be expended over five years. This amounted to a per capita rate of a little over six pounds per year. The ten million was spent but the proposed railways, this time from Cloncurry to the Gulf and the Via Recta from Ipswich to Warwick didn't eventuate. The money ran out. The lines that were completed cost much more than the budget estimates. There was nothing left. While there was a change in Premier from McIlwraith to Griffith, the representation for Gympie also changed. In Gympie, Bill Smyth, who also had a strong mining background, represented his home town while Dr. Jack who continued to ally himself very strongly with the Northerners, successfully contested the seat of Cook.

William Pattison, a butcher by trade but by good fortune one of the owners of Mt. Morgan's mountain of gold, also made his debut into Parliament. Another new parliamentarian was Donald Wallace from Clermont, whose main claim to fame was as owner of mighty Carbine, the superhorse, Melbourne Cup winner and inspiration for the naming of the mining township, Mt. Carbine, in north Queensland. While Dr. Jack was still mining in the Rockhampton area, he often met Wallace when he came in from his Clermont property. They had been classmates together in Melbourne, at Charles Goslett's Academy and at Scotch College. Wallace thought Dr. Jack was wasting his talents trying to make a living from mining and repeatedly urged him to 'give up the life of a goldseeker and return to his friends.' Now they could be friends together again.The Ninth Parliament saw the entry of another of Jack's old friends, Robert Philp, a staunch Northerner, as the Member for Musgrave. It also saw the entry of Isador Lissner from Charters Towers. Lissner was a small-statured man with a big heart and an equally significant sense of humour. In his maiden speech he confessed he was a poor speaker but a 'very good list'ner'.[72]

The 1883 election campaign in Cooktown was anything but clearcut and straightforward. October 30th was the date set for the election but owing to bushfires in the Cooktown hinterland results from 'Cooktown, Middle Oakey Creek, Daintree River, Deep Creek, Halpin's (Harvey's Creek), Byerstown, Fountainville (Cannibal Creek), California Gully (head of the Tate River), Normanby Diggings, M'Ivor River, Mitchell River (Callaghan and Wilson's place), Bloomfield River (Bauer's Plantation) and Lower Laura River (near Jones's)' could not be telegraphed. Cook was entitled to double representation. Four candidates nominated, Messrs Campbell, Fred Cooper (barrister), a grazier Lumley Hill and Dr. Jack Hamilton. Campbell and Dr. Jack spent a few days

72. C. A. Bernays, *Queensland Politics During Sixty Years 1859-1919*, Govt. Printer Brisbane, p101

campaigning in Cairns before leaving for Cooktown. At both places, Jack was met by enthusiastic supporters who welcomed his return. Cooper and Lumley Hill followed later to Cairns.

By October 31st, thanks to the Telegraph Stations, returns were in from all but the thirteen Cook hinterland booths. Campbell, polling his best at Watsonville, a tin mining area out from Mt. Garnet, was slightly ahead of Dr. Jack with Lumley Hill and Cooper level in third place. When the missing thirteen booths' returns came in the score was reversed. Hamilton (839) and Cooper (767) were duly declared. However, the results from Halpin's and California Gully were astounding. It was alleged that at the latter polling place there were only fourteen resident voters but Jack and Cooper secured 178 votes each, Lumley Hill 23 and Campbell, 2. At Halpin's twenty-five voters secured Dr. Jack a further 50 votes, Cooper 43, Campbell 14 and Lumley Hill 7. The Brisbane *Courier's* correspondent noted that, 'There is great excitement owing to (Jack Hamilton's) 50 votes polled at Halpin's and 180 at California Gully. It is stated there are not more than 50 voters in both places together.'

The port of Cairns was a staunch McIlwraith stronghold and Dr. Jack and Cooper polled well there. They left from Cooktown on the first boat after the election to take up their seats in the new parliament. No one could explain what could have caused the discrepancies in the Halpin's /California Gully polling as the scrutineers advised that a great deal of liquor was consumed throughout the day.

The north didn't have it all on its own. Paddy Perkins from Aubigny was charged with bribery and corruption in connection with the same election. He was not found guilty but the Aubigny election was determined null and void. Perkins was the Secretary for Public Lands at the time of the great railway debate. He was also the owner of a very large brewery (now Castlemaine Perkins) the end-product of which was so freely distributed, especially among the railway navvies whose vote Perkins was courting, that a contemporary wrote it 'would fill the dry bed of the Flinders River out Hughenden way.' As any Hughendenite knows, it takes an awful lot of liquid to do that. Paddy Perkins lost his seat. Another Campbell, James, secured it in March 1884. Later Paddy won a seat in the Upper House, the Legislative Council. B.B.Moreton (Burnett) also attracted a petition against his election in the same poll.

In January 1884 Lumley Hill and Campbell petitioned against Hamilton and Cooper. Dr. Jack stood his ground and let the inquiry decide. He retained his seat and Campbell ousted Cooper. Naturally the Cairns *Post* took a keen interest. In the issue of March 6th 1884 it wrote, 'the mere fact of knowing for certain who our representatives are carries a sort of consolation with it for being deprived of a member whose politics were after our own heart. The Cook has not been fortunate in its recent attempts to elect legislators and has pursued a course quite original and different to the rest of the colony. First of all the polling was postponed until after all the other elections, because it found there were so few voters on the roll. When this slight oversight had been rectified and two members returned in the Conservative interest, the Liberals suddenly alleged there were too many voters who had polled.'

The Election and Qualifications Committee 'in their Solomonlike wisdom decided to divide the honours. The Conservative Hamilton and the Liberal Campbell were given the seat.' Fred Cooper 'refused to interfere in the petition.' and 'has entirely shaken off from his boots the dust of Cook; that, in fact, he purposed obliterating from his mind all recollection of any of its inhabitants.' The *Post* was of the opinion he would not be greatly missed.

On the other hand 'Hamilton is a tower of strength in himself and the conduct of Mr. Hamilton stands out in bold relief to his credit when compared to Mr. Cooper. Regardless of an attempt to unseat him and whilst the Committee was sitting, he boldly and fearlessly spoke out in condemnation of the policy of the Government not withstanding that the committee was composed almost entirely of Government supporters. He has never lost an opportunity of furthering the interests of Cook or making our wants known to parliament. Mr. Campbell is as yet an untried man and new to parliamentary life so will have to prove himself before any opinion can be expressed of his worth and capacity. We have no doubt, however, that although he will sit opposite to his fellow member on all party questions, whenever the interests of the Cook are in question, there will be no diversities of opinion between them'.

Despite being Cairns' hero, Jack didn't win all his battles on their behalf. On the same page is a paragraph saying his 'endeavours to procure some Government assistance for the damage done to the Herberton Road and the Barron Bridge by the late severe weather, have proved futile and that the Government refuse to make any grant whatever.' Jack was at that time in the Opposition.

Even in Herberton, where Campbell polled well, the correspondent wrote, after a paragraph on the continuous and never-ending rain, that 'in default of a whole loaf (two Conservatives) you have half a one; and in my opinion, a very big half too. Mr. Hamilton will stir up your interests with due alacrity, that's certain and I fancy Mr. Campbell will give no reason to be complained of. As reasonable men and gentlemen, they will drop their political differences when the requirements of Cook are considered.'

One of Cook's requirements was the railway to the Palmer. Bashford's tender was accepted for the first section and work was to begin as soon as the weather permitted. The cost was for a little over 100,000 pounds and already Cooktown residents were 'wearing a brighter aspect than they have done for many years.' Bashford also nominated for Cook in the 1883 elections but withdrew his nomination when he heard of Dr. Jack's switch from Gympie. Cook, like Gympie, remained eternally loyal to Jack. The Hodgkinson was the one exception. One wonders why.

ROBERT CHRISTISON AND THE POOLE ISLAND FREEZING WORKS

1883 was one of the years Mother Nature sought revenge on mankind. When the island of Krakatoa between Java and Sumatra disappeared in a volcanic explosion it took the entire population of 36,000 with it. A much lesser act of destruction was the cyclone off the north Queensland coast that destroyed Robert Christison's pride and joy, the Poole Island freezing works. Christison of Lammermoor, himself a cattleman, saw the urgent need for a freezing works along the coast. Unlike most of his fellow cattlemen he decided to do something about it. When he initiated the scheme he found very willing supporters amongst the seemingly apathetic so that by 1881 he'd sold all of the 1500 shares on offer. The finance raised purchased the site at Poole Island and fortune was still with Christison when he secured contracts for 6000 bullocks annually for five years at a lower price than he'd expected.[73]

With a start like that he'd have to tempt the Fates. He must have, for they retaliated. The machinery for the freezers was so badly designed and constructed, that it had to be re-built. Christison, the practical man, saw the deficiencies as soon as the machinery was installed. He told his wife that the shock nearly turned his hair grey on the spot. It was later found the so-called engineer was mentally deranged. He tried to drown Christison and some of his men as they rowed back from a tour of inspection. They survived the attempt but the perpetrator took laudanum and died despite all endeavours to rush him to Bowen by steam launch.

While waiting for a replacement engineer from the manufacturer, Christison tinkered around until he 'rectified the error' and got the machinery 'going at last'.Because of the faulty machinery the works lost out on a contract to London for the 'largest meat cargo yet carried in one ship'. The *Sorrento*, which was to have taken on frozen meat from Poole Island, had to fill its freezers in New Zealand with the record load – or go home empty. That wasn't the end of Poole Island's problems. The meat that should have been sold frozen had to be rendered down so that when a ship, the *Chyebasa* did call, it took on 'not quarters of frozen beef but casks of tallow'. The manufacturers finally fixed the ailing machinery but not

73. G. Bolton, *A Thousand Miles Away,* Jacarandah Brisbane, 1963 p101

before the *Fiado* was forced to be diverted to secure loading at rival Lake's Creek freezers and the *Sorrento*, headed a second time for Poole Island, struck a leak and foundered off Monte Video. The *Fiado* didn't escape the Fates' wrath either. Fever broke out on board delaying her arrival in Rockhampton by two weeks in which time a fire at Lake's Creek destroyed the meat she should have been carrying. Meat processing, like cattle raising, was a chancy business.[74]

By January 1884 the Fates still hadn't finished with Christison's freezing works. Everything looked rosy. The freezers were working at top efficiency, the *Fiado* was berthed ready to load at first light. Success seemed finally assured until – the cyclone struck. Nothing but havoc and devastation reigned. Even the monumental masonry wall designed to protect the reservoir was completely disintegrated. 'Seabirds were killed. The *Fiado* parted her cable and was driven on to the mainland. Trees snapped like carrots'.

Financially Christison's company was ruined. They had no way of raising the 5000 pounds needed for restoration. The British India Co. took over but after a few shipments faded from sight. Christison, an incurable optimist, remarked, 'Poole Island has opened the way for other ventures and we shall yet see export factories studded all over Australia'.

DR. JACK'S PARLIAMENTARY DEFENCE OF WARDEN HODGKINSON ON THE PALMER GOLDFIELD

About this time Jack took up the cudgels for Warden W.O. Hodgkinson in Parliament when Hodgkinson was accused of 'plagiarising' a 'report purporting to be a Report of the Palmer Gold Field' in late 1883, allegedly colouring the report to attract investment. Hodgkinson asked for an official inquiry to clear his name but his application was ignored. He was an ardent Griffith supporter but Griffith didn't reciprocate in this instance. He then requested Dr. Jack to take it up in Parliament. Hodgkinson, an ex-journalist, entered the mining industry as a young man in Charters Towers. He later entered Parliament and as Griffith's Secretary for Mines was responsible, among other things, for the New Gold Fields Act which extended the 'newness' of the field to three years thus effectively keeping Chinese miners out for that length of time.

Dr. Jack made scathing comments about the members selected to investigate Hodgkinson.[75] He could not 'fancy the Hon. Member for North Brisbane going up there and taking the direction and dip of the reefs, and the direction of the underlie, and the character of the rock!' Griffith was the Hon. Member referred to. Hamilton thought the matter best left to someone who understood the

74. M. M. Bennett, *Christison of Lammermoor,* Alston Rivers London, 1927 p161
75. *Votes and Proceedings,* Govt. Printer Brisbane

context, the Government Geologist. He was sure Hodgkinson would be exonerated in a fair and informed inquiry. Part of the problem he considered, was that Hodgkinson had 'spoken in favour of some claims the Hon. Minister for Mines had stated Sir Thomas McIlwraith was interested in; and they all knew the name of Sir Thomas McIlwraith was like a red rag to a bull when mentioned before the Minister for Mines.' Dr. Jack had checked the list of shareholders of the mines in question and McIlwraith's name was not among them 'The reefs were the 'Comet', the 'Queen' and the 'Ida'. Hodgkinson told of the Queen reef crushing over two ounces to the ton for three and a half thousand tons of ore until it came upon a belt of poor stone owing to the intrusion of a sandstone bar.' The only way the committee could find the truth or otherwise of these statements was to consult the 'mining registrars upon the Palmer.' 'The statements,' Hamilton emphasised 'were true'. He 'was in a position to know that as well as anyone in the House, for he had been six or seven years in the Queen reef, though he was not interested in it now, and could have no motive for speaking well of it.' Hodgkinson's report summed up by saying that 'the average crushing per ton from more than a hundred claims in the area yielded over two ounces'. 'The great gold-producing colony of Victoria cannot exhibit an average of over half-an-ounce per ton.' Clearly Dr. Jack felt that both his friend the warden and his goldfield were being victimised.

By the time Dr. Jack's defence of Hodgkinson was in progress, the Queensland National Bank foreclosed and took over the Queen. Hamilton borrowed to add two new boilers encased in stonework to supplement the original boiler and also to install a five stamp battery. This done, the sandstone intrusion was encountered and the gold production decreased below a sustainable level. It was worked on tribute for a while and then in 1881 a new company took over, obtaining 414 ounces in their first year. By 1887 the flooding was so bad the mine could no longer be worked. Half a century later the battery that had been Jack's pride was taken to a mine near Cooktown. But none of the blame for this or any other financial failures could be laid at Hodgkinson's door. It was the way mining went. Some prospered. Some walked off.

The Northerners stuck together even though they lost the skirmish that day. Eventually, by the loyal efforts of his friends, Hodgkinson was completely exonerated. Today, Jack Hamilton would most likely be termed a 'maverick' but it was against his nature to stand aside while justice was being denied, even though the Minister for Mines and the Minister for Works both stated in the debate that they 'cared nothing' for his opinion on the matter.

DR. JACK'S PARLIAMENTARY DEFENCE OF CHRISTIE PALMERSTON

Hodgkinson wasn't the only friend Dr. Jack was called upon to support in Parliament. His contemporaries were right when they said he could be a relentless enemy at times but the staunchest of friends to his mates. The Pathfinder, Christie Palmerston was in trouble with the Magistrate Walsh at Geraldton (Innisfail). Christie had been taken to court by a Chinese miner Lee Cook who alleged Palmerston had assaulted him then looted or burnt his belongings. Walsh took a rather one-sided view of the case and disregarded Palmerston's counterclaim that Lee Cook had stolen 40 ounces of gold that was later recovered from some bushes. Walsh also discounted Palmerston's four witnesses who claimed Palmerston was with them on the day Lee Cook claimed he assaulted him and destroyed his camp.

Palmerston discovered gold on the Russell River inland from Geraldton. It was not a major field but some miners went there. A second field was discovered by a Chinese miner who then had the right to work his claim. This discovery was not as rich as the first location but Chinese flocked to it in their hundreds and overflowed from there to the first site. The opinion held by the European miners trying to make a living was that while the discoverer was entitled to operate his mine his countrymen were bound by the new Act to wait three years before they could legally enter the field.

The Magistrate reported that the Chinese miners were getting extremely poor returns but the miners insisted that thousands of ounces of gold were being smuggled out of the country. Gold in Hongkong brought almost twice the price offered for Russell River gold in Queensland so that gold smuggling was a highly profitable undertaking. Walsh chose to ignore any argument brought up by Palmerston and his supporters and concentrated on Palmerston's business dealing with the Chinese, providing safe conduct and beef for their consumption. He bound Palmerston over 'to keep the peace for three months' with two sureties of fifty pounds and a hundred pound bond from Palmerston himself. Payable immediately. With gold from the Russell valued at not much over three pounds an ounce it came to a heavy imposition.

Feeling he was being treated unfairly Palmerston wrote to two parliamentary friends Henry Palmer, another northerner who came into the Eighth Parliament and Dr. Jack.[76] In those times backbenchers were permitted to move for the adjournment of the House when business for the day was concluded and bring up matters of their own. Jack tabled letters he had received from Palmerston and

76. Paul Savage, *Christie Palmerston Explorer*, J. C. U., 1989 p233

other miners on the Russell complaining of the magistrate's bias. 'It is not for me to suggest what action the Government should take in this matter; but it is evident serious charges have been made against the police magistrate, and some inquiry should be made as to their truth or untruth, especially when they are made by a man whose word can be relied upon. Many members of this House know Mr. Palmerston personally as a respectable truth-telling person, who could have no possible interest in making these charges, and who wishes to be allowed an opportunity of proving them. Many others in this House who may not know him personally may have read with interest an account of his explorations in Northern Queensland which appeared in the *Queenslander* and other papers. He is considered the best bushman and explorer in North Queensland. I have omitted a number of serious statements against the police magistrate, because I do not think it exactly fair to repeat them; I have said enough to show that some inquiry should be made.'

Dr. R.L.Jack and Jim Mulligan supported Palmerston's summary of the worth of the Russell Field also but the magistrate kept up his vendetta. A few years later Palmerston left his new wife and baby daughter to go exploring and prospecting for gold near Kuala Pilah in the Malay Peninsula. Here he died January 15th 1897. He died not, wrote his employer to his widow, 'of any special disease, but from the effect of 30 years of the hard life he had lived'. Palmerston was 45.[77]

THE BURNING OF PREMIER GRIFFITHS' EFFIGY IN FAR NORTH QUEENSLAND

Griffith, the smooth-speaking lawyer, as Premier, was eager to unite the colonies into a Federation. In Ayr angry residents unhappy with his stance on Kanaka labour burnt his effigy in protest. The citizens of Port Douglas also burnt his effigy when Port Douglas lost out to Cairns as the outlet for the Herberton tinfields and terminus to the proposed railway. Federation was of lesser importance to Northerners at that time. They wanted a separate northern colony as their first priority. Federation could follow. McIlwraith, too, was not sure if the timing were right for Federation. His 'vision' was for a strong, independent Queensland well able to hold its own at the bargaining table with the more powerful colonies before a Federation was formed.

Griffith's dream won out and in 1886, the Federal Council of Australasia met for the first time and included delegates from Queensland, Tasmania, Victoria and Western Australia. New South Wales and South Australia did not put in an

77. Paul Savage, *Christie Palmerston Explorer,* J. C. U., 1989 p266

appearance but the little colony of Fiji was represented there by the Colonial Secretary, the Hon. W. MacGregor.

Griffith also brought in a Members' Expenses Act, the first in the Colonies, which made provision for a maximum of 200 pounds available to members annually with a 'deduction for every day absent'. The Act was knocked back twice by the more affluent members of the Upper House but eventually, after some clever ploys initiated by Griffith, they capitulated. Travel expenses at one shilling and sixpence a mile were also allowed. Dr. Jack was very relieved at this outcome.

The payments came too late for Willy Rea.

7

CAPITAL, UNIONS, PATRIOTISM AND THE ADVENT OF THE 1890s DEPRESSION

Unsuccessful moves for Separation of the North.Melanesian labour. Repeal of McIlwraith's Railway Companies Preliminary Act by Griffith. Russian scare. Cannon and two rifles sent to Cooktown. Qld Navy and Defence Force. Migration encouraged. Griffith-Dutton Land Act. Corfield enters Parliament. Drought. Griffilwraith coalition. M.P's salaries raised. Bill of Rights. Slump. Recession. Tick and pleuro epidemics. M.P's salaries halved. Maritime Strike. Shearers' Strike. Death of Macrossan. Brisbane floods. Jack's swim.

MACROSSAN AND DR. JACK FIGHT FOR A SEPARATE STATE FOR FAR NORTH QUEENSLAND

Over the years Jack's Parliamentary interests widened. Goldmining matters possibly took precedence and his knowledge was obvious in the cases he put for his miner mates in Mine Regulation Bills dealing with overall mine safety, security of ownership and in trying to update various Goldfield Act Amendments. He took up the cause of railway development and, as well as promoting rail links in the north, he heartily supported a survey for a rail line from Brisbane to Gympie. He was there to join in the cheering when the railway from Cairns to Herberton was officially opened. Jack understood from personal experience the difficulties isolation brought with it and he spoke earnestly and often in favour of the construction of roads, tramways and bridges in the bush.

On 20th August 1886 Macrossan introduced his first resolution regarding self-government for the North. Dr. Jack was one of the Northern Nine who championed both him and his cause. One of the more impassioned speakers, Jack was none the less supported just as fervently by fellow northerners Philp, Chubb, Henry Palmer and Isidor Lissner. Macrossan outlined the history of the movement. He told how the Duke of Newcastle had twice[78] recommended self-government for the northern part of the Colony in Queensland's infancy. With passion he dealt with the benefits Separation would bring to the top half of the Colony and backed up his words with irrefutable and impressive statistics. Parliamentary recorder Charles Bernays wrote, 'Those who remember the fervour he put into his speech; the care with which he marshalled his facts; his detailed statements of the reasons for severance, and the piles of statistics with which he backed up his argument could not but believe in the justness of the cause and the earnestness of the man.'

Samuel Griffiths spoke in an equally impassioned manner against Macrossan's motion to petition Her Majesty 'to cause the Northern portion of the Colony to be erected into a separate and independent Colony with representative institutions'. Griffith was concerned about the use of South Sea Islander labour in the far north and the effect that it would have on the proposed Colony. He was backed up, very garrulously, by William Brooks whom Bernays considered a 'monomaniac' on that subject. Dr. Jack, Robert Philp, Henry Palmer, C.E. Chubb (who came into the Eighth Parliament with Jack and rose to become Attorney General) and Isidor Lissner from Charters Towers spoke strongly in defence of Separation but, after a lengthy fourteen days of discussion and debate, only the Northern Nine were in favour. The motion was lost by forty votes to nine. W.H.Corfield, the western pastoralist who temporarily turned goldrush teamster before coming into the Tenth Parliament had, the previous year to Macrossan's speech, made representations to the Secretary of State to have 19,000 square miles of his Winton electorate included within the boundaries of the hoped-for new northern colony. The Wintonites, too, were disappointed.

PREMIER GRIFFITHS LEGISLATES AGAINST CHINESE IMMIGRATION

Griffiths passed an Act limiting the introduction of South Sea Islanders in December 31st 1890. It was bitterly contested. Melanesians were brought in during the 1860s when Queensland had visions of becoming a cotton-growing superpower. Cheap coloured labour was deemed a necessary ingredient for

78. C. A. Bernays, *Queensland Politics During Sixty Years 1859-1919,* Govt. Printer Brisbane, p516

success. It was thought Europeans and people of European descent could not stand up to hard physical work in the Tropics. Some of the Palmer's early hard-rock miners could easily refute this idealogy. Louis Hope's sugar-growing experiments led to sugar eclipsing cotton in the plantation owners' dreams. Governor Bowen encouraged the use of Melanesian labour providing there were safe guards to protect their well-being. Dr. Jack, like many Northerners and supporters of McIlwraith, agreed with him and consistently spoke in support of that labour in debates in the House, always provided the rights of the Islanders were considered at all times. Abuses of the system were far from rare though they weren't rife and acts were often passed to control breaches of humanitarian conduct.

While the plantation owners, and through them, most northern politicians, vociferously supported the use of Island labour, an equally vocal opposition espoused the cause of Anti-coloured Labour. Many of these were the Liberals led by Griffiths with William Brooks the most vocal spokesman. The greatest abuse of Island labour occurred in the 'blackbirding' ships. The plantation owners were relatively benign and many of the kanakas looked upon the canefields as their second home. However, in 1872[79] a law was passed to 'prevent and punish criminal outrages upon Natives of Islands in the Pacific Ocean'. Those who drew up the legislation didn't mince words. They called their law the Kidnapping Act of 1872. The maximum penalty for breaches was 'the highest punishment other than capital punishment' in use in the Colony where the offender was tried.

Select Committees 'looked into' the practices involved and found that in the majority of cases there was no cause for complaint. By an 1880 Act[80] there was an age limit on 'recruits' and shipmasters were required to lodge a quite considerable bond redeemable only when the labourers were returned to their native islands. Medical and hospital treatment was to be provided at all times; a wage was set; the ration and clothing allowance was increased and the practice of deducting store accounts from wages was banned. Employers could use Melanesians only in tropical or sub-tropical agriculture such as sugar, cotton, coffee, tea, rice, spice and tropical or sub-tropical fruit

'A perusal of this Act does not convey the impression that the Government of Queensland in 1880 was a consenting party to a slave trade, nor does any sane person believe it was,' Bernays commented.[81]

79. C. A. Bernays, *Queensland Politics During Sixty Years 1859-1919*, Govt. Printer Brisbane, p67
80. C. A. Bernays, Op Cit, p68
81. C. A. Bernays, Op Cit, p68

During Griffith's regime, the Act was amended to stipulate that work in tropical agriculture meant 'fieldwork' and not the work as performed by engine-drivers, blacksmiths, wheelwrights, grooms or coachmen, horse driving or carting unless in the course of field work duties. Domestic or household service was forbidden. By protecting the Islanders, Griffith was also protecting the jobs of white workers.

With Melanesia under close scrutiny Griffith was appalled to find shipowners were by-passing the usual South Sea Islands and were recruiting from the Louisade Group to the north where the islanders had very little understanding of the implications of their so-called 'agreements'. For the illicit forays of the *Hopeful* in which ship the northerner Robert Philp had an interest, the death sentence and life imprisonment were brought in to punish 'murder and kidnapping'. The capital sentences were never enacted.

Islanders who had been resident in Queensland for five years prior to the Act and who did not wish to return to their native islands, were exempted by the law to repatriate Melanesians by the end of 1890. The five year lag would also enable the planters to become accustomed to the new labour conditions. As Islanders were brought in on a three year contract this really meant an adjustment period of eight years. Later, in 1892, Griffith reneged on his decision and repealed his former Act. The years from 1892 to Federation in 1901 were decidedly mute on the subject but after Federation those Islanders not exempted under the previous Act were unceremoniously returned to the islands.

MACROSSAN'S LAWS AGAINST CHINESE IMMIGRATION

Robert Philp of the Burns Philp merchantile empire came into the Ninth Parliament in 1883. He was later, in a McIlwraith ministry, the Minister for Mines and Works. Here his commercial experience, Scottish canniness and modest, unassuming demeanour stood him in good stead. He was a kindly man and a loyal friend as Dr. Jack was to find when his mining income slumped into the red. Another Northerner and McIlwraith man, John Murtagh Macrossan, introduced a Chinese Immigrants Regulation Act which restricted the number of Chinese passengers arriving on any boat to 'one in every fifty tons register'. In addition, a landing fee of thirty pounds was imposed and the combination of the two succeeded in 'limiting this form of Asian migration'. The Parliament also repealed the *Labourers from British India Act* which would have allowed coolies to be brought in to work on plantations and pastoral holdings.

The Act whose repeal brought the most pleasure to Griffith and his Liberals was The Railway Companies Preliminary Act under which McIlwraith had been trying to set up his ambitious rail network. The repeal successfully brought to an abrupt end the negotiations which had been proceeding with the former Government.

Dr. Jack's parliamentary duties kept him busy. He, Macrossan and Philp were active in promoting the northern part of the colony, improving communications, encouraging mining exploration and generally airing the complaints of their remote constituents. Hamilton had small interests in mines other than the Queen and in land but investment – or speculation – was a risky business in a new colony. The Queen was a constant worry to Dr. Jack. There were signs that, due to water and a depleting supply of gold, it would soon be worthless. A party was working it on tribute but it was highly capitalised and the returns to all were poor.Adding salt to the miners' wounds were stories that Chinese working the alluvial in Gregory Gully, just outside the Queen's boundary, had found several very large nuggets.

Jack was also finding that in a more civilised world his medical skills were proscribed by his lack of the necessary registration. Until the Payment to Members Act his parliamentary career was becoming a luxury he could not afford. In 1885 there was a postscript to the Halpins Gully/California Creek episode. Tom Campbell, Dr. Jack's parliamentary partner, had enough of trying to run the colony and resigned. His place was taken by one of the men he ousted, Lumley Hill, who won the ensuing by-election. In 1887 there was a redistribution and though Townsville and Charters Towers retained dual member representation, Cook was reduced to one member. Fortunately for Dr. Jack, he was that member. The output from the Gympie mines was declining and as yet there was no shining star of discovery in the furtherest North.

COOKTOWN GETS AN 1803 CANNON AND TWO RIFLES FOR DEFENCE AGAINST THE RUSSIANS

In the early eighties there was a Russian scare, the nineteenth century equivalent to Reds under the Bed. Germany was also flexing her muscles, showing more signs of wanting to increase her colonial power. Though Britain did finally acknowledge McIlwraith's foresight in his annexation of part of New Guinea, British naval and military chiefs considered the Colonies should do more for their own defence and not merely rely on a Britain who might not be able to come to their aid. Cooktown residents felt especially vulnerable. If either the Russians or the Germans converged with ill intent on their peaceful township they'd need guns to repel them. They asked for guns. Dr. Jack passed the message on and the

Government generously sent Cooktown a cannon of 1803 vintage and two rifles. Fortunately they were not needed in the defence of the settlement.

Griffith, while the Member for North Brisbane and Minister for Works, was eager to equip a defence force and money was allocated during McIlwraith's stint as Premier to purchase two gunboats, one for the Thursday Island area and one to be stationed in Moreton Bay. They were the *Gayundah* and the *Paluma* symbolically named for the indigenous words for thunder and lightning. They were sister ships, under 130 feet in length and with a ten feet draft. With a top speed of ten knots they were quite ferocious with 20 cm,15 cm, and two 4 cm guns plus two machine guns each. The Colonial ships were entitled only to fly the Blue Ensign but when Griffith pledged the *Gayundah* to serve with the Royal Navy Squadron when it was in Australian waters, she became the first Colonial ship to 'wear the White Ensign'.[82] For years, the two ships carried out very useful coastal surveys but saw very little active warfare.

Carried away with defence, Queensland also ordered a torpedo boat, the *Mosquito*, which was shipped out to Brisbane. The *Mosquito* was the first torpedo boat for the newly-formed Queensland Marine Defence Force manned mainly by volunteer sailors. With a top speed of 21 knots and a capability of firing a 36 cm torpedo, she was soon followed by a converted steam piquet boat, the *Midge*, which also fitted with torpedo gear. The Queensland Navy was determined to put a bit of bite and sting into its fleet. Until the advent of the two torpedo boats, the army was responsible for torpedoes which were, at that time, mines laid on the sea floor and exploded by controls from the shore. The *Gayundah*, besides its White Ensign, could lay claim to another first. It made use of wireless telegraphy.

The Army was upgraded and the Queensland Mounted Infantry became the more familiar Queensland Light Horse. Unfortunately, while all eyes were searching for a possible human enemy, an animal one slipped in, unnoticed for a brief time, from below the border. The rabbit plague arrived in its superfluity from the southern colonies and work was begun almost immediately on the world-famous Rabbit Fence.

In the southern states workers were forming unions, notably the wharf labourers and the shearers. William Lane, later to become well known for his effort to form a Utopian, socialistic society, New Australia, in Paraguay in the nineties, started his left wing paper the *Boomerang* which led to the establishment of *The Worker* later to become the mouthpiece for the as yet unconceived Labour Party.

---

82. Cilento and Lack, *Triumph in the Tropics*, Smith and Paterson, Brisbane, 1959 p319

William Lane's Utopia. Dr. Jack Champions the Poor Battlers

With thoughts of his early mentor, Dunmore Lang, on his mind, Dr. Jack objected to immigration officers being sent to London and European centres to encourage migration to the colony. None were being sent to Scotland. 'No doubt', he said, 'the experience the Premier has had of Mr. Thomas McIlwraith as leader of the Opposition has led him to think that there were enough Scotchmen in the colony already.' He also objected strongly to an honourable member remarking that 'there was neither money nor brains in the North' saying whoever made that statement was 'deficient in brains' and that both Charters Towers and the Palmer had out-performed anything the South had to offer. A contemporary once said of Jack Hamilton that, in more mundane matters of Parliament he was so quiet as to be almost mute but when he espoused the cause of something he believed in, he was both eloquent and persuasive. His dry wit surfaced when feelings ran high. During the debate on the Triennial Parliaments Bill (which he did not support, preferring five year terms) many emotional outbursts occurred. In reply to one caustic attack Dr. Jack stated amiably that he 'thought the Hon. Member for Fassifern was mistaken in saying the Hon. Members lost their tempers; they were all in a most seraphic frame of mind'. He did not personally care how long the Committee (on Triennial Parliaments) sat. Having missed his train, it'd be far more comfortable to stay all night in the warm chamber than to go hunting for a bed.

He spoke with an insider's knowledge on the plight of miners, both those working their own claims and those working for others. Begging to disagree with a statement made by another mining man, Stubley, that 'claim-owners had always a very great regard for the safety of the men who were working for them', Dr. Jack believed that 'it was not the case'. The strong took advantage of the weak. He was still the man who supported the underdog and tried his utmost for their protection in the Mines Regulation Bill within the realms of 'commonsense practicalities'. Despite his unswerving loyalties, his head was never in the clouds.

In a Crown Lands Bill he was again on the side of the battler and pointed out that if the country wanted migrants, it would have to compete with other countries such as the United States of America, to secure the very best. Comparing the policies of Queensland and the U.S.A. under the old Bill, Queensland paid the migrant's fare but once arrived, the 160 acre land grants had to be paid for, albeit over a period of five years, at a cost of twenty pounds. In America, the immigrant paid his own fare – four to five pounds for a single man – but land was granted free and the new landholder 'meets with no vexatious restrictions whatever'. In

Queensland, under the new Bill, the migrant would be required to lease the land, fence it and generally improve it so that, at the end of the ten years he has spent in building up the asset, he still has to pay a pound an acre to get title to it. 'It is immaterial to the tenant whether the landlord is a State or a private landlord. What is of importance to him is a fair landlord'. Suburban land was sold with an upset price of a pound an acre without the necessity to expend any money or labour upon it before being granted title. Should the rural landholder strike bad seasons or setbacks that prevented him from paying the annual rent, he stood to lose the block plus all capital and labour expended on it.

Jack Hamilton didn't restrict his concern to the constituents of his own electorate. While Member for Cook he supported the claims of his old Gympie electorate for bridges over flood-prone streams. He championed the Croydonites over sundry grievances concerning the lack of communications and the constant postponing of work on the Normanton to Croydon railway. He pushed for roads in the Cook district and was equally determined on the survey for a Herberton to Georgetown rail link. His approach to problems associated with prospecting on Pastoral and Grazing Farm Leaseholds was even-handed and practical. He didn't support the Suppression of Gambling Bill but neither did the majority of the electors who voted for him.

Despite the start of the Cairns to Herberton railway in 1886 at which ceremony Premier Griffith, watched by the Member for Cook and other Northerners, turned the first sod, the North was still concerned about the Brisbane Government's disregard for the needs of the top half of the colony. Maurice Hume Black, a sugar planter from Mackay, and Isidor Lissner sailed to Britain to put the case for Separation yet again to Sir Henry Holland, the Secretary for State. And yet again, that 17th May 1887[83] Sir Henry and Her Majesty's Government didn't think the Northerners' arguments were sufficiently strong for their wish to be granted.

During his term as Premier, Griffith did his best to redress the issues of pastoral leases and lands which had been held in reserve against the building of McIlwraith's proposed railways. In the Griffith-Dutton Land Act he aimed to encourage closer settlement by restricting the pre-emptive rights of large pastoral lessees. Resumed areas were divided into smaller holdings, 'grazing farms', and offered in areas of up to 20,000 acres for a thirty year lease. In the arable lands, Agricultural Farms of up to 1280 acres were surveyed with a lease of 50 years and the option of converting to freehold title after ten (later reduced to five) years. Dr.

83. C. A. Bernays, *Queensland Politics During Sixty Years 1859-1919,* Govt. Printer Brisbane, p517

Jack approved of these moves and voted accordingly but unfortunately this well-intentioned Act and Griffith's Ten Million Loan led not only to a boom and unbounded optimism but also to unhealthy land speculation and inflated prices.[84]

What railways were built, some thousand miles of them, were paying for themselves but it was a period of high immigration, doubtless a valuable long term asset but a serious drain on the immediate public purse. With rail and telegraph expansion, as well as growth in the private sector, the value of imports exceeded money brought into the colony by exports. They had a budget deficit and were over-spending.

Education was also upgraded. Public Libraries expanded to over 60,000 volumes for some 5,000 subscribers. Progress was phenomenal but it came at a price. The Griffith Ministry recorded four successive deficits despite bringing in levies of 5% (to rise the following year) on goods as diverse as machinery and timber. The duty on spirits increased to the equivalent of 25c per litre but even these impositions had little effect on the demand on Treasury to pay its bills.

The Treasurer, James Dickson – later to become Premier – resigned as Colonial Treasurer in a reaction to the new Land Tax which he considered favoured leaseholders above freeholders.[85] In the election of June 1888 for the Tenth Parliament, McIlwraith stood in the two-member seat of North Brisbane with his arch rival but erstwhile ally, Sam Griffith. Dr. Jack's friend and one-time patient from the Palmer days, W.H. Corfield, came into Parliament as did his old Gympie associate Horace Tozer and a newer one from Cairns, Fred Wimble of the *Cairns Post* and Wimble's Inks. One of the issues dear to Fred's heart was the establishment of reserves for Aborigines. He presented a Bill to Parliament which was successful but fellow Cairnsite, Archie Meston, who was appointed Protector of Aborigines, liked to claim full credit for the idea. Yarrabah was eventually founded near Cairns by the Anglican Church in 1892 and populated with Aborigines from near and very far.

PREMIER MCILWAITH AND WHIP DR. JACK AGAINST IMPERIAL GOVERNOR MUSGRAVE

The fore-runner of the modern Labour Party, Tom Glassey, who came to Australia only four years earlier, gained a seat at this election. Glassey helped organise the coal miners' union at Bundamba and initially was not at all popular in a Parliament of mainly conservative representation. Most of his fellow

84. C. A. Bernays, *Queensland Politics During Sixty Years 1859-1919,* Govt. Printer Brisbane, p103
85. Murphy and Joyce eds., *Queensland Parliamentary Portraits,* U. Q. P., 1978 p168

members mellowed towards him. He was a moderate and a great fan of Griffith's but McIlwraith's attitude towards him softened very little. The new Member for Bourke, John 'Plumper' Hoolan, the editor of the Georgetown newspaper the *Mundic Miner*, was also a radical. So forthright was the *Mundic Miner* that allowances for libel payments were included in the paper's normal running expenses.

Later Glassey quarrelled with the Labour Movement but initially he was the fore-runner both of the party and its platform of righting the wrongs perpetrated against the working man. Scornful of the discrepancy in wages paid to the powerful and the 'masses' he once remarked that no man was worth five hundred pounds a year but when he left the Legislative Assembly as Queensland's representative in the Federal Senate, it was at a salary of six hundred pounds a year. Bernays noted, 'there is no record of his having paid back to the Treasury the extra hundred pounds that he said he was not worth'.

McIlwraith headed the Tenth Parliament from June 1888. Despite his independent stance, Jack became McIlwraith's Whip. At the end of November McIlwraith resigned over what superficially seemed a trivial case – the Kitt Case. Benjamin Kitt stole two pairs of boots valued at two pounds and was sentenced to three years imprisonment. The Parliament recommended the suspension of the jail sentence as Kitt was previously of good character. Instead the conditions of the Offenders' Probation Act would apply. Governor Musgrave exercised his authority over the elected Parliament and refused to accept their recommendation. The Executive again sought Kitt's release by Royal Prerogative from Her Majesty's representative and again Musgrave refused. Naturally, McIlwraith used his firestick approach when he might have done better with a more conciliatory request but here he saw yet another case – only five years after the refusal to recognise his government's annexation of New Guinea –where an appointed Governor outranked an elected Premier. Where was the 'responsibility' of an elected government if their wishes could be so easily ignored?

Possibly Musgrave thought the Kitt Case might have been a devious ploy to obtain a precedent for the impending *Hopeful* black-birding case.

McIlwraith resigned over the Governor's refusal but withdrew his resignation when Musgrave received a belated authority from the Secretary of State to free Kitt. This time, Samuel Griffith sided with McIlwraith. 'In practice the Governor should not, even although his own opinion may differ from that of his Ministers, refuse to follow their advice unless' 1. Imperial interests are involved. 2. The offence is against the laws of the Empire. 3. It is an abuse of the Prerogative.[86]

Griffith, the barrister, could not see that any of these exceptions applied and that there was no reason at all for McIlwraith to stand down.

Both the Governor and McIlwraith were in poor health. McIlwraith suffered from peripheral neuritis, an excruciatingly painful disease probably caused by an imbalance of whisky in an inadequate diet. There is no evidence that it affected his brain but his temper did not improve with his affliction nor the continual expense it brought him for no relief of his suffering. The public brawling took its toll on both participants. Musgrave died 'within the month' and was replaced by the popular General Sir Henry Wylie Norman and McIlwraith relinquished the premiership to Morehead while retaining a place in Cabinet.

The Depression of the 1890s and the Rise of the Unions

Money for railways was brought up again. Horace Tozer spoke for eight hours straight on the subject of a million pounds loan for the same and earnt for himself the nickname, 'Jawbone'.[87] McIlwraith resigned yet again in a huff but came back when he thought the storm was over to form with Griffith the Griffilwraith Cabinet of August 1890. He was secure in the knowledge that, as Colonial Treasurer, he still held the real power.[88] Another *Payment of Members Act* was passed to Jack's relief raising members' salaries to 300 pounds a year with a mileage allowance for travelling.

With gold becoming increasingly hard to find miners were keen to exploit any possibility of finding the precious metal even if it was to be done at the detriment to others. Dr. Jack moved a resolution to prevent mining under streets, 'that no mining lease of streets on a goldfield be granted unless the power to grant the right of passing under such streets was retained.'[89] The debate of this interesting motion continued without accomplishing anything and gave Dr. Jack little choice but to withdraw it. Away from Parliament he found himself obliged to transfer his mining interests from gold to the baser metals of tin, wolfram, cassiterite (stream tin) and copper. Stream tin was to be found in abundance in California Creek, the site of his phenomenal electoral success and Queensland soon became the world's top producer of this metal. Copper was discovered by Pat Molloy at Mt. Molloy and at Zillmanton, Calcifer and Chillagoe. Silver and silverlead were treated at Montalbion, Stannary Hills and Muldiva. Herberton was the main centre for this area and here Dr.Jack purchased a residential block, a *pied de terre* for his northern

86. C. A. Bernays, *Queensland Politics During Sixty Years 1859–1919,* Govt. Printer Brisbane, p118
87. C. A. Bernays, Op Cit, p120
88. C. A. Bernays, Op Cit, p120
89. C. A. Bernays, Op Cit, p370

visits. Smelters were introduced, used extensively and removed to other centres as the supply dwindled. When the *Quetta* sank in the Torres Straits with the distressing loss of 133 lives, she was carrying silverlead concentrate from the Totley silvermine.[90] Small gold finds were reported, coal deposits found in the Cooktown area, but no potential Palmer Rivers were uncovered. From Parliament, Dr. Jack was granted two thousand pounds for gold exploration and this was used prospecting the northern half of the Peninsula.

Griffith introduced something like a Bill of Rights but did not proceed with it. It proposed the 'Equal right of all persons to life and freedom of opportunity'. [91] Another principle stated 'The natural and proper measure of wages is such a sum as is a fair immediate recompense for the labour for which they are paid but it can never be taken at less a sum than such is sufficient to maintain the labourer and his family in a state of health and reasonable comfort.' The duty of the State was 'to make provision by positive law for securing proper distribution of the net profits of labour in accordance with the principles (28 in all) hereby declared.' It was considered outrageously radical in those days of a hundred years ago but it seems reasonable enough now. Griffith put a high value on education and introduced small schools into isolated mining camps to cater for families there. He inaugurated travelling schools in trains and pioneered what developed into a correspondence school for excessively isolated children.

The nationwide slump came early to North Queensland which relied so much on gold and sugar exports. Land prices tumbled. A Cooktown firm which purchased land in Cairns in 1886 for 7,000 pounds found four years later that it was valued at less than half that amount and there were only very reluctant buyers at that price. The same applied in Townsville where Flinders Street frontages leased at fifteen pounds a yard of frontage, but only two years later could not find takers at ten shillings a yard. Pastoral values dropped accordingly and poorly producing mines were worthless. The dreaded cattle tick moved in to compound the disaster and a fatal bovine disease, pleuropneumonia, was in epidemic proportions in beef herds and dairies. Working bullocks were not immune to the pleuro and the ticks which brought with them a deadly fever, 'Red Water' so-called because in the advanced stages, the urine from the crippled kidneys was almost pure blood. Only horse teams operated with any strength and regularity.

An independent lot unused to unemployment, workers laid off in the North soon took the lead of their more southern comrades and formed unions in an

---

90. N. Q. Register, *North Queensland's Rich Mining and Exploration History 1865-1998* p9
91. C. A. Bernays, *Queensland Politics During Sixty Years 1859-1919*, Govt. Printer Brisbane, p121

attempt to combat the slump. In Charters Towers, noted even more for its independence of spirit, the movement was exceptionally strong. Labourers, wharfies, navvies from the railways, butchers, bakers, shop assistants and even typographers and women formed unions. Plumper Hoolan and Dunsford, the latter a one-time owner of a struggling newsagency and gift shop, broke up with the help of a crowd of their followers, a meeting of the Charters Towers Separation League. They considered that any new northern colony would be pro-kanaka and jobs would be even harder to find. Republicanism was rife. The British and their Imperialism bore the bulk of the blame for the depression. It was their banks that cut off the credit.

Dunsford came into the Eleventh Parliament and in 1897 was responsible for a Mining Commission set up by Philp after a letter from Dunsford. W.H.B.O'Connell chaired the commission with Anderson Dawson (later to become the first Queensland Labour Premier), John Hoolan, Bill Smyth and 'Dr' Jack Hamilton.[92]

Dry years added to the countryman's worries and through his lack of solvency the smaller townships suffered.

Mid-1890 saw the Maritime Strike which paralysed for a time both exports and imports. The strikers met one of their first reverses in quiet little Cooktown when striking wharfies tried to prevent the merchants unloading their goods. The townspeople, totally dependent on the ship's cargo for supplies, unexpectedly turned on the strikers who had up till then gathered a good degree of sympathy. This time they came off second best. The unrest at this time set the scene for the lengthy Shearers' Strike which began just months after the birth of the Griffilwraith Government, the coalition formed from expediency when neither Griffith nor McIlwraith could muster the numbers to go alone.

Caught up in Federation plans and dreams, Griffith was out of the colony for most of the time and McIlwraith's poor health forced him to leave the affairs of state to others. The chief of these was Horace Tozer who became the man of the moment on the Conservative side. Quite possibly his pomposity did little to ease the tensions between the 'workers' and the 'capitalists' despite efforts of more conciliatory men. Parliament's sympathies were not with the striking workers. They had only two supporters in the House, 'Plumper' Hoolan and Tom Glassey. Hoolan was the Wild Man of politics, extremely eccentric and fond of 'impolite

92. C. A. Bernays, *Queensland Politics During Sixty Years 1859–1919*, Govt. Printer Brisbane, p121

adjectives'.[93] The Irish-born Glassey, quietly spoken and moderate in all things, was almost his complete antithesis.

In 1891 Griffith took Macrossan with him as a co-delegate to the Federation council meeting in Sydney. Although they were political opponents Griffith had a personal liking for Macrossan with great respect for his integrity and his ability. As Harrison Bryan says in his biography of Macrossan, the 'intensity of his conviction and the detail of his thought combined with the remarkable extent to his knowledge of constitutional law and practice to illumine a speech…'.[94] Macrossan, rarely in the best of health, was overtaken by a debilitating illness. He made his last speech supporting Griffith on his ideas for the Senate on March 17. On March 29 'he took to his bed and at four o'clock next day he had gone'.[95] Macrossan's dream was for the day when people thought of themselves as Australians first, the designation of Queenslander, Victorian etc. taking a secondary place. Jack the Hatter died alone and in a strange city.

The resumption of land for Grazing and Agricultural Farms cut almost a third of their area from some of the biggest runs and rents were increased markedly on the others. The pastoralists were not happy. Wool prices were down, the economy depressed and rainfall well below average. All this came on top of the ticks, pleuro and rabbits.

Unionism continued to gain strength in Queensland and the other colonies. Until 1886 unions had been illegal. William Lane was expounding on the virtues of unionism in his paper and the message was spreading. Townsville had a Trade and Labour Council and Tom Glassey toured the bush with union officials whenever he saw the opportunity, seeking active memberships. By 1890 the Australian Labour Federation had over 20,000 members. The A.L.F. wanted to nationalise all 'means of producing and exchanging wealth' with monetary compensation paid and to provide pensions for children, invalids and the aged – those unable to earn their own livings. Creditable aims but the owners of the money-making assets didn't think so.

THE LEAD-UP TO THE SHEARERS' STRIKE OF 1891

Shearers were paid according to the number of sheep they shore. The rates ranged from the current equivalent of from $1.75 to $2 per 100. In the closer settled southern colonies where the expense of getting from shed to shed was

93. C. A. Bernays, *Queensland Politics During Sixty Years 1859–1919,* Govt. Printer Brisbane, p133
94. Harrison Bryan, *John Murtagh Macrossan* (*Queensland Parliamentary Portraits.* Murphy and Joyce eds), U. Q. P., 1978 p109
95. Harrison Bryan, Op Cit, p117

negligible, the rate tended to be lower.[96] A shearer using hand-activated blades shore from 80-90 sheep in a normal day. At the top rate this gave him today's equivalent of around $9.50 for a five and a half day week. From this he had to pay for the upkeep and replacement of his shears, which, if the wool were full of grit and sand –quite common in dry years – could be a considerable sum. He had to pay for his tucker, usually today's $2 a week, as shearers had healthy appetites and appreciated good food. The station-owner in turn provided meat (mutton) and sometimes vegetables, eggs and milk. The average shearer probably netted about $7 per week. This was slightly more than the politicians now received as remuneration. Their newly-gained salaries were cut by half in the recession.

If a sheep were damaged or killed during shearing, owners often deducted the price of the animal's worth from the shearer's pay. There was no Workers' Compensation or Social Security so shearers were often called to contribute when a mate was injured or sick. If the workers needed to purchase anything while at a shed, they usually had no option but to use the station store where mark-ups could be very high. As shearing was not a year-round occupation, many of the men often worked in other industries in the 'off' season. A *Brisbane Courier* reporter surveying the men at the Barcaldine strike camp found he was unable to separate the shearers from the labourers present as half of them held tickets in both unions. Twenty-one weeks work was a fair annual average for shearers in 1890.

Stationhands were paid less, today's $2 a week, but they had continuous employment and were supplied with bed and board. Shearing, especially the old blade shearing, is particularly hard work inducing repetitive strain injuries to the blade hand and enough back problems to fill a textbook on the subject. Wet sheep were abhorred and detested as shearers considered the wet fleece gave them fevers and accelerated the onset of arthritis. They refused to shear sheep if they were 'wet'. In more recent years it was said that 'wet' sheep for the Kiwi shearers who worked the Australian sheds just meant that the water was too deep for them to bend down to grasp the shears without drowning. They didn't have 'wet' sheep in New Zealand.

Gun shearers made a very good living although it was often made at the expense of their health. Jackie Howe, who shore 321 sheep near Blackall at Alice Downs in just 8 hours and 40 minutes, was later able to buy a pub and become a capitalist.[97]

96. Stuart Svendsen, *The Shearers' War*, U. Q. P., 1989 p43
97. Stuart Svendsen, Op Cit, p45

'Agreements' were brought in to govern shearers' conditions and pay but this led only to more trouble. Non-union labour was introduced and was violently opposed by the unionists. Union camps were set up and rioting broke out between union and non-union men and, of course, the military who were sent to 'keep the peace'. Many of the shearers were ex-miners who were forced to look for alternative work when the gold cut out. William Browne, a little younger than Dr. Jack, gained his mining experience at Gympie, then moved on to the Palmer and Herberton before settling at Croydon. Here he was secretary and later president of the Miners' Union. He lost an eye in an accident before he stood against Plumper Hoolan for Bourke. Hoolan won that round but when Croydon was made a separate electorate, Browne won the new seat.

Another miner-turned-shearer was the gentlemanly William Hamilton who was sentenced for his part in the 'Shearers War' to two years hard labour on St. Helena Island in Moreton Bay. He and thirteen others were convicted of no less than twenty counts of 'conspiracy'. Griffith, as Attorney General, commented that 'conspiracy was the most elastic offence known to law'. It could stretch to cover anything. William Hamilton was a calming influence on the rioters who created mayhem, burnt out paddocks or burnt down shearing sheds, at least once with sheep penned inside. His influence was so moderating that the law and the military tried their hardest to avoid arresting him and, when he was finally apprehended, the local military commander, Jackson, is supposed to have offered a 2000 pounds cash bond to bail him.[98] Even at his trial, three Crown witnesses, volunteered character references for him but he declined their efforts to help his case. He felt he could not desert his mates.

MORE BANKS FAIL AND DR. JACK SWIMS THE FLOODED BRISBANE RIVER

Like many Queenslanders, Dr. Jack's feelings were mixed. He shrank from the bullying tactics of some of the wealthy pastoralists but neither could he condone the behaviour of some of the extreme troublemakers among the 'underdog' shearers. It was a distressing time for all – lightened by a few carefree escapades such as when the soldiers, shooting and barbecueing emus for a change of diet, decorated their slouch hats with what was to become the Light Horseman's badge of honour, his tuft of emu feathers.

French, who was in charge of the military, was a rather bombastic man who confronted the ragged shearers decked out in a uniform so weighted down with gold braid that one ex-miner striker mused aloud to his mates, 'Wonder how

98. Stuart Svendsen, *The Shearers' War,* U. Q. P., p165

many weights (penny weights, a gold measure) he'd go to the dish?' Jack's personal problems gave him enough headaches without taking sides in the dispute. His mine was abandoned, his property worthless and he was declared insolvent. To add insult to injury, this came at the time member's salaries and expenses were reduced to 150 pounds per year. All the while, travel expenses to and from his remote electorate remained unchanged. Dr. Jack was an 'influential backbencher' and tried to have the salaries re-instated but his influence didn't reach as far as the Treasurer or Treasury. He was fortunate to have in Parliament friends whose losses hadn't been so absolute. Even bluff, outspoken McIlwraith was known never to turn a blind eye to a friend in need.

Gold was discovered at Starcke, a little to the north of Cooktown but Dr. Jack's hopes of his mate John Dickie finding an El Dorado further north weren't realised. Perhaps his old schoolmate Donald Wallace gave him the tip to back Carbine in that year's 1890 Melbourne Cup.

The Shearers' Strike was still unsettled in 1893 when more banks failed, throwing even deeper gloom over the colony. The Brisbane River rose twice in three weeks in record floods which took eleven lives.[99] Victoria Bridge and the Indooroopilly rail bridge were destroyed, the city inundated and three ships stranded in the Botanical Gardens. A Reverend Stewart argued that the floods were the 'wrath of God against the prevalent desecration of the Sabbath'.[100] The sight of the flooded river roused memories in the old Northerners of the turbulent Palmer and how their Dr. Jack swam it at its height to rescue Paddy Shanahan. Jack seemed to have an affinity with water. After his success in the 1888 election, the *Cooktown Courier* noted that 'instead of shouting, Mr. Hamilton invited them (his supporters) to come for a swim in the river'. Crocodiles aside, there are other ways to cool off on a tropical August day besides wetting the tongue in the bar. Donald Wallace boasted of Jack's swimming prowess while at Scotch College and how he held an underwater swimming record which was still, at that time, unbroken.

Challenges were made that Jack couldn't repeat his Palmer performance of almost twenty years earlier. He couldn't let a southern river beat him. 'Make it worth his while and I wager he'll do it,' suggested one of his northern mates and the bets were laid. The crossing decided upon was below the Victoria Bridge in the

99. Cilento and Lack, *Triumph in the Tropics,* Smith and Paterson Brisbane, 1959 p347
100. *From the Moreton Bay Courier to the Courier Mail 1846-1992* p78

vicinity of the Southside Cultural Centre, opposite Elizabeth Street. Despite his fifty plus years, Jack made the crossing and collected the jackpot.

ℬ

8

THE RISE OF LABOUR IN QUEENSLAND POLITICS

Increased Labour membership of Parliament. Anderson Dawson. Andrew Fisher. Q.N.Bank crash. Jack insolvent. William Lane to Paraguay. Jack Government whip. Tattersalls Affair. Jack attempts to have members' salaries reinstated. Three steam trains a week from Cooktown to Laura. Griffith put forward case for three autonomous provinces. Rejected by Legislative Council. School of Mines, Charters Towers. Separation of North Queensland passed in late night sitting. Dr. Roth. Cyclone Mahina destroys pearling fleet. Dawson first Labour Premier. Philp Premier. Increased Stamp Duties. Lost Government. Jack lost his seat to Labour.

CRASH OF THE QUEENSLAND NATIONAL BANK AND INSOLVENCY OF DR. JACK

The Eleventh Parliament of 1893 had increased Labour representation, 16 members in all. Andrew Fisher, a one-time Scottish coalminer, now Member for Gympie, was one of the new Labour men. He continued his political career in Federal politics, where in 1910, he led the world's first Labour National Government. A by-election in January 1894 elected a seventeenth Labour member, an Independent Labour candidate, Ogden[101] He had already won the seat of Townsville in the by-election following Macrossan's untimely death in March 1891. Ogden missed out on Labour Party endorsement for the 1893 elections when he insisted on electoral reform. Another Labour entry was Anderson Dawson, ex-miner and journalist from Charters Towers. His

101. Christine Doran, *Separatism in Townsville,* J. C. U., 1981 p66-67

mining job was that of 'amalgamator', a position of paramount trust. He was in charge of the amalgam plates which collected the pure gold processed from the crushed ore. His constituents held him in the same high trust his ex-employers showed towards him. Six years later he, too, made Labour history. He became Queensland's first Labour Premier, a position he held from December 1st to December 7th. A period of six days.

Griffith resigned to become Chief Justice with a salary especially upgraded to 3,500 pounds to attract him. McIlwraith pulled the Colony out of a cataclysmic financial mess when the Queensland National Bank crashed and seven of the remaining banks suspended trading. Despite the end result, McIlwraith's reputation was under a cloud of suspicion over the financial collapse and though he was exonerated of any impropriety, his health, never robust, again failed. In his efforts to save the colonial economy he was helped considerably by his pastoralist friend James Tyson who bought half a million pounds worth of Government bonds.[102] This generous act was of no small assistance in saving the Colony from Jack's individual fate of ignominious insolvency. Tyson's good deed earned him a seat in the Upper House, the Legislative Council. He held the seat for five years during which time he made one speech but even that, he complained, was misreported in the media.

With politicians' salaries still halved, the Premier was looking for further economies when he noticed the Sergeant at Arms, an 'elderly gentleman who sat at ease at the bar of the House.'[103] His formal attire was 'knee breeches, silk stockings, silver-buckled shoes and an official cut-away coat'. The Premier thought it would be a prudent housekeeping measure to combine the office of Sergeant at Arms with that of the Clerk Assistant and thus save the Treasury 300 pounds a year. That done, in October McIlwraith wiped his hands of the problems and again resigned the Premiership. He took leave because of ill health and went to Europe seeking treatment. Nelson became Premier.

William Lane took time off from writing up the shearers' case to ship on the *Royal Tar* with 220 members of the New Australia Co-operative Settlement for Paraguay. He might have been better advised to sail to Western Australia where Pat Hannan had just discovered Kalgoorlie's golden treasure chest. T.J.Ryan tried to end the Shearers' Strike with a Peace Preservation Bill which in Parliament did not have the slightest pacifying effect. During the consideration of the Act, seven members, Messrs Browne, Reid, McDonald, Dawson, Turley, Dunsford and

---

102. C. A. Bernays, *Queensland Politics During Sixty Years 1859-1919,* Govt. Printer Brisbane, p133
103. C. A. Bernays, Op Cit, p134

Glassey, were suspended and removed from the chamber.[104] The Bill for the Preservation of Peace was passed the following year and the terms were so draconian that Premier Nelson equipped it with safeguards so that it couldn't be brought into operation without the surety of certain representations to the Governor.

JACK THE WHIP AND THE TATTERSALLS SCANDAL IN QUEENSLAND PARLIAMENT

Jack was now Government Whip, a recognition of his place in the ranks of Parliamentary long-termers. The supporters of the Labour opposition were also making changes. The Australian Workers Union was formed by the amalgamation of the Shearers and Shedhands Unions in eastern Australia. Not to be outdone by progress on the mainland, Tasmania introduced flat-rate taxation. In South Australia, David Shearer stole some of their thunder by developing a steam-powered car and South Australia's Parliament, equally innovative, gave women the right to vote.

For some time George Adams of Tattersalls had been trying to set himself up in Queensland. Proprietor of the Tattersalls Hotel in Sydney, he organised the first of his 'sweepstake consultations' in 1881 and regularly ran lotteries until the New South Wales Government, in a righteous frame of mind, closed his business twelve years later. He moved up to Brisbane, setting up at the Telegraph Chambers in Queen Street. From there he ran a personal and mail order sweepstake business until public opinion went against him and the Queensland Government Gazette of September 1st 1894 published a notice directing that any letter addressed to 'Tattersall c/- Geo Adams, Queens St., Brisbane should not be registered, transmitted or delivered to him'.[105]

Jack Hamilton took up the case for the gambling man and asked the House to cancel the order made in the Gazette. It was passed by 28 votes to 9. At the end of the sitting, a happy and victorious Jack issued a general invitation to the members as they filed out, 'Come on, boys. It's Adams' shout!' This led, naturally, to the assumption that Jack had been paid by Adams to look after his interests. A thousand pounds was the most popular amount floated. *The Worker* took up the matter with uninhibited glee.

'The Lord prosper all our Consultations is a sentence in the prayer read at the opening of each sitting of Parliament, and by way of variation, Jack Hamilton ran a small consultation on his own account the other evening on the ballot for a

104. C. A. Bernays, *Queensland Politics During Sixty Years 1859-1919,* Govt. Printer Brisbane, p136
105. C. A. Bernays, Op Cit, p137

Select Committee. Tozer, who is just after introducing an Anti-Gambling Bill, is reported to have taken a bob's worth in his sweep. The novelty of the Assembly Chambers being turned into a tote shop was really a very comical satire on the efforts now being made to suppress the gambling evil'.

Of the three Brisbane newspapers, *Brisbane Courier, The Worker* and *The Telegraph* only the latter supported Adams, although on December 14th 1895, *The Worker* accepted and printed an advertisement for Adams' consultation just a few pages over from their report of his agent's dastardly deeds to subvert the integrity of both the paper and its readership. According to Editor Higgs, a Mr. Bird who had some connection with *The Telegraph* approached him to print a leader in favour of Mr. Hamilton's Consultation Bill (Jack's efforts to have Tatts and other lotteries given legal standing). Bird was willing to pay five pounds, the going rate for the writing of the leader, plus advertising rates. Higgs said he was too busy to do anything just then but asked Bird to return at 8:30 that evening.

Bird returned and unknown to him, two of Higgs' associates were closeted in the adjoining room to eavesdrop as witnesses. Bird brought a list of suggestions Jack had written and given to him as an outline for the article. With parliament sitting and he being Whip, Jack didn't have time to write it himself. He estimated the lottery would return 30,000 pounds annually in Government duties and provide employment for a hundred people as well which, he argued when accused of bribe-taking, was reward enough in his consideration. Bird thought the Bill would go through if the Government 'do not make it a party question'. Higgs didn't consider that five pounds for writing the article, even if it were paid for at advertising rates, was a 'very large sum to offer a man to go back down from his principles'. After all, Adams was reported to be making 100,000 pounds a year. A few thousand wouldn't be much to get the Bill passed. Bird later told the Select Committee Higgs wanted the 'same as the Members are getting in the House. 25 pounds'. Bird didn't think that was right or that it could be done. Five pounds was the usual rate for writing a leader.

Seymour, one of Higgs' witnesses and sub-editor of *The Worker* thought he'd heard enough to justify the headline for his article, 'Bribery and Corruption'. The war of words continued in *The Worker* and *Courier* all the while the Select Committee deliberated. Jack attended the hearings, presumably as witness to part of it but was criticised by *The Worker* for taking notes during the evidence given by Higgs, Seymour and Higgs' other witness, Hinchcliffe. Higgs considered Jack should have had the 'good taste' to stay away. The Committee found that the statement that Higgs was offered money to write the leading article in support of the Consultation Regulation Bill wasn't contradicted and this might be taken as some justification for the article his newspaper published on bribery and corruption. However, the evidence given by the editor of the *Brisbane Courier* did not disclose any reasonable grounds for the article that appeared in his newspaper. The charges made by the *Courier* and *The Worker* alleging Members acted dishonourably were not true. The Committee did not recommend any further action.

Jack admitted asking fellow Parliamentarians who were also lawyers to draw up his Private Member's Bill but that was because he wanted to be sure the wording was correct. Quite understandable given the circumstances, but he needn't have bothered. Correct wording or not, on a Bill that would have given the lotteries legal standing, the votes were 38 to 9 but this time they were against the lotteries. The Gambling Bill failed and George Adams moved on, first to Hobart where he stayed for over fifty years and then, in 1954, to Melbourne.

Adams was earlier a driver for Cobb and Co. and rose to hold the Road Manager's position. Well known among his fellow reinsmen as a more than competent driver, he often took the reins for a stage when travelling on his old employers' coaches. Adams' money is said to have carried the Tattersall Curse. Three of its beneficiaries certainly haven't prospered in recent times. Jeremy and Vincent 'John' Adams became news when brother John was imprisoned in India without trial for nearly two years and Jeremy's plane disappeared upon a flight to Flinders Island. The man charged with the Port Arthur massacre was identified as the heir to a fortune bequeathed by Adams' partner, David Harvey – money made from Tatts consultations.

(An interesting snippet in the column beside the one in *The Worker* reporting the Select Committee's ruling read '14 officials belonging to the Sultan of Turkey's household, suspected of treason, all die on the same day.' Who thought history and research were dull?)

DROUGHT, FAR NORTH QUEENSLAND RAILWAYS AND A NEW MINING ACT

Jack had no luck either with another Private Members Bill trying to restore parliamentarians' salaries and expenses to the old level. The Upper House exerting its privilege, refused to ratify the Bill on previous occasions also. Jack estimated the cost of the exercise would be 6,400 pounds for the year. *The Worker*[106] considered the Member for Cook was acting in an extremely selfish and greedy manner 'to preserve personal ends'. A fair proportion of members refuted

106. *The Worker,* Brisbane, 8th Dec. 1894

this accusation but time to pass the Appropriation Bill that would enable the payment was running out and there were more important matters to discuss. Jack couldn't get permission for an adjournment granted so that he could put the amendment to re-instate the salaries and the Appropriation Bill was passed with a majority of one. The close vote nearly brought the Government down and led *The Workers'* poet-in-residence to write some caustic verses on A Majority of One.

What did cheer Jack considerably was the news from the North. William Lakeland had discovered good but not outstanding gold at Rocky Creek near Coen; Joe Baird found the Mt. Carbine wolfram deposits and, at long last, the construction of a rail line from Mareeba to Chillagoe and Mungana was begun. The Cooktown line was not only through to Laura but was running three steam trains a week. Some 14,000 passengers made use of the service in 1888 but less than ten years later, with the demise of the Palmer goldfield and no substantial new claims to the north, it was being called a White Elephant by its detractors. It progressed no closer to its Maytown destination than an impressive bridge just on Laura's outskirts. The only train that used it was the engine that performed the ceremonial crossing at the opening festivities.

With Federation waiting in the wings, Separation took a back seat but Griffith put forward an alternative suggestion for three autonomous provinces, North, Central and South Queensland.[107] This caused a split in the Separatist camp. Some, tiring of knocking their heads against the Colonial Office's brick wall of intransigence, decided to accept Griffith's scheme but the Legislative Council rejected the provincial alternative as well as the separatists' case. Lord Ripon wrote to the Governor, Sir Henry Norman, that 'in view of the present financial and commercial position of the Colony, the moment does not appear to be favourable for any scheme or schemes of territorial partition'. Disillusioned, the Separatists reluctantly abandoned the public espousal of their cause.

In addition to the financial and political upheavals which disturbed the rural calm, the colony was still in the grip of a drought which extended through into the twentieth century. It was 1903 before the cycle of low rainfall was broken. Pastoralists were thankful for their artesian and sub-artesian bores put down on the geological guidance of Robert Logan Jack. Without them, there would have been no water for the sheep- and cattle- men or for their stock. The small town of Thargomindah had further cause to bless their underground water supply. Their town was lit by a hydro-electric system powered by the local bore. It was Australia's and possibly the world's, first hydro-powered settlement.

---

107. C. A. Bernays, *Queensland Politics During Sixty Years 1859-1919*, Govt. Printer Brisbane, p519

Jack, like many other miners and ex-miners, was absorbed by the news of the Klondyke rush in America. There was still enough of the adventurer in him to wish to join in but years of experience tempered the appeal a new strike once had for him, especially one halfway across the world. Shields and Johnson located some gold on a creek near Coen and named it Klondyke in honour of the big rush but it never lived up to the expectations of the name bestowed upon it. A Mining Commission was appointed by Philp to make a study of the industry and to come up with a Mining Act to suit the current times. W.H.B.O'Connell, who came into McIlwraith's Tenth Parliament and was later Minister for Lands, was appointed to chair the commission. Jack was one of the four member team, all from the Legislative Assembly and with practical mining experience. Anderson Dawson of Charters Towers, John 'Plumper' Hoolan and Bill Smyth, the Gympie miner, who took the seat of Gympie when Jack returned north to represent Cook, made up the committee.

Jack was pleased when the commission's research work took him back to his old stamping grounds. Many of his mining mates had gone but there were still others to voice their grievances and to offer suggestions to benefit the comprehensive Act that was put to Parliament the following year. Philp introduced the Act and was greatly helped, not only by his commission members but also by the assistance of warden Philip Sellheim, a very competent mining man. It wasn't all plain sailing. Philp himself spoke 170 times in defence of his Act and when some Labour members requested that the Bill be withdrawn he lost his characteristic patience. 'There are only nine or ten mining members who thoroughly understand the Bill. Let them discuss it and let everyone else look on and listen to what they say.'[108]

Fortunately they accepted his advice and it became the most up-to-date and streamlined mining legislation in Australia. Included in it were policy suggestions for the establishment of Schools of Mines and one was opened in Charters Towers two years later. Some time previous to this Act, Philp brought in an amendment to the old Goldfields Act to provide commons for domestic stock on the goldfields and to levy a per head charge on stock pastured on them. What prompted him was an exceptional case of the owner of a fifty acre homestead block running several thousand head of cattle on the goldfield while contributing only the minimal rent for his small homestead lease.

108. G. C. Bolton, *Robert Philp* (*Queensland Parliamentary Portraits* Murphy and Joyce eds), U. Q. P., 1978 p204

The same year the Commission investigated the mining industry, there occurred in Parliament what Charles Bernays in his book *Queensland Politics During Sixty Years* calls 'the Remarkable Occurrence of November 1897'.[109] Just when the Separatists had given up – in fact G.S.Curtis, the leader of the Central Queensland Separationist Party, left Brisbane to go back to his electorate- the matter of Separation was not only brought up but carried on the casting vote of Speaker Cowley. The euphoria it induced in the Northerners was fleeting. Next day, twenty-five irate members not with Northern interests, informed Governor Nelson by letter that the result was not representative of the whole Parliament. The resolution was carried in 'a very thin House, after a late sitting the night before'. They wished to record their 'emphatic dissent from the terms of the resolution'. They had been, if only by chance,out-manoeuvred but unfortunately it did the Separationists little good. There were no changes to the boundaries but late-night Parliamentary attendances rose accordingly for some time.

JACK HAMILTONS OPPOSITION TO ROTH AS CHIEF PROTECTOR OF ABORIGINES

While McIlwraith led, Jack was a supporter although he did not always follow the party line if he were particularly concerned about a course of action. His first duty, he contended, was to the people who elected him and not to maintain his side's lead in the popularity stakes. He was wearying of parliamentary life and yearned to get back to mining interests but he had unfinished business to complete. He had no personal difference with the Chief Protector Roth but 'rather', as Jack said, 'a personal bias in his favour. Once Roth walked 14 or 15 miles to vote for me.'[110] On the other hand his sense of justice was affronted and he took exception to the libellous and unfounded charges Roth made in a current report about some of Jack's Cook constituents and to Roth's neglect of his real duty of improving the lot of the indigenous in the Far North. Repeatedly Jack tried to bring up in Parliament the unworkability of the Act and the need to amend it but he was ridiculed, not only by his arch-enemy *The Worker* but by some of his fellow members. His requests were ruled 'Not Formal' and he was not allowed to proceed with them.

Roth was a medical doctor as well as an anthropologist and Dr. Jack often considered that he often put the latter ahead of the former profession in his dealings with the Aborigines. This certainly went against Jack's sense of commitment to his fellow men. Requests for Roth to visit natives who, though

109. C. A. Bernays, *Queensland Politics During Sixty Years 1859-1919,* Govt. Printer Brisbane, p530

110. *Votes and Proceedings,* 29th Oct. 1902

being treated with whatever medicines and skills were available, would undoubtedly improve much faster under treatment from a qualified doctor equipped with better drugs, went unheeded. In one case, an eighteen year old Aborigine was quartered but '400 yards' from Roth's hotel. Roth decided against paying the patient a visit and he subsequently died but Roth decided against visiting the large Aboriginal camps at Burketown and Normanton for the opposite reason, 'because there was no sickness there'.[111]

Cooktown held a public meeting to support an amendment to the Act controlling Islanders and the indigenous people. The permit system was particularly vexatious to employers with isolated bases. Some prospective employers, a good proportion of whom were bêche de mer boat-owners, could not get permits to employ natives despite having excellent character references and no record of abuse. Others were given work permits for six months while the lucky ones had permits for the full year. According to the meeting, the discrimination was inexplicable. Communications were difficult and the bêche de mer unpredictable. Owners often found that, just as the bêche de mer were 'running', it was time to return to Cooktown to apply for a new permit and thus risk missing the catch. Restrictions also applied to the areas from which owners could 'recruit', although many had recruited from the prohibited areas for years before the ban was brought in.

Boat owners preferred mainland Aborigines to the imported Islanders and Japanese. One boat was crewed entirely by women divers who took an unholy pride in out-performing their male compatriots. Quite probably the ship's captain had no permit for these gems as 'unmarried men', which included a good proportion of the bêche de mer catchers, were not permitted to employ females. Some form of agreement was used at times in place of a permit but as the signatories were usually limited in literacy skills, these were often open to misunderstandings. It was considered that the men from the South Sea Islands understood the contractual arrangements well enough for practical purposes but that the men from the northern islands were less knowledgeable.

Burns Philp boats and stores were discriminated against because of the *Hopeful* case, though Robert Philp at this time had little to do with Burns Philp policies. He was requested some time previously to stand down from the Board because the number of shares he then held were too low to qualify him for membership on the Board of the company he'd helped develop. Proprietors of public houses, many of whom were Labour supporters and hence fiercely anti-Philp though they

111. *Votes and Proceedings*, 30th October 1893 (1 a.m.)

had to purchase their grog supplies from the company, found that they, too, could not obtain permits to employ.

Unless permits were acquired, white families who had reared Aboriginal children from an early age were liable to be charged with 'harboring' and the children could be taken away, either to a Mission or in some cases given in marriage to an Aboriginal who made legal application through the Protector. This caused considerable anguish when the bride was happily, though illegally, 'married' to a stockman of her choice from the same station.

Through Dr. Jack, the Cooktown meeting sent a 539 word telegram – the only means of reasonably fast communication – to the Premier and 'for which' Jack's detractors alleged, 'the State had to pay'. Billy Browne from Croydon stood back from the heat of the debate but agreed judiciously that, 'if the Act is wrong, amend it'. Vincent Bernard Joseph Lesina, Joe to his friends, came into the 13th Parliament as Member for Clermont. In this round, he was certainly not on Jack's side.

'Dr. Roth has raised about him a host of enemies and the man who could produce so many strenuous enemies must be a man of particularly good character – a man who was fearless in administrating the Act in the interests of the black people of the Colony.'

Despite the opposition, Jack persisted in his request for an inquiry. He also came into possession of some photographs Dr. Roth had taken of naked 'male and female Aborigines in the most indecent positions'. Natives were suspicious of having any photographs taken, more especially photos like the ones Dr. Jack referred to. Jack also alleged that Roth used his official standing to obtain artefacts which were sold overseas or to Australian museums. In particular, Dr. Roth obtained skulls and bones, desecrating a burial site in the name of science and overlooking marked faults in the employer's care of the natives under his control in return for the opportunity to obtain the relics.

One special case Dr. Jack took to heart was of a young girl of mixed blood, Lizzie Johnson, who was taken from her adoptive father (who wanted to marry her when she came of age!) and placed at Yarrabah, the Mission recently set up on the Cairns Inlet. Jack threatened to resign if Lizzie and others like her weren't returned to their homes. Gleefully the Cairns *Morning Post*, which even-handedly criticised both Roth and Yarrabah, printed a rhymed contribution from 'Booyongs' to 'celebrate a recent stiff fight in Queensland parliament over' Lizzie.[112] Part of it is as follows:

112. *Morning Post Cairns,* 16th Dec. 1902

'Jack was awful wroth, of course,
But still he held the fort,
And swore a Government like this
Would not get his support.
To vote for such a government
He simply would decline,
As he couldn't join the Labour crowd,
Then of course he must resign...
...And the scene 'tis said was awful
When Jack and Foxton met
But Jack is getting reconciled.
He ain't resigned; not yet.'

The Hon.Colonel Julian Fox Greenlaw Foxton C.M.G. was Philp's Home Secretary having earlier been Nelson's Minister for Public Lands. He was an ardent reformist, either repeatedly trying for, or bringing on, laws like Women's Suffrage and other voting reforms. The Factory Act Law was one he managed to have brought in to govern the rules and obligations of employment. In answer to Dr. Jack's allegations about Dr. Roth, Foxton was of the opinion that 'Dr. Roth was a scientist and that his work in this connection was recognised not only in Cooktown but throughout the world. The Australian Aborigine most closely resembled the neolithic and paleolithic types of man and therefore they were unique from a scientific point of view. They also had certain rites and customs, information with regard to which was desired by scientists in all parts of the world. It was purely in the interests of science that these photgraphs were taken'.

To which Jack caustically replied, 'Science was at a very low ebb if an officer like Dr. Roth was allowed to coerce Aborigines in order to procure photographs of them in indecent positions'.

He wanted an inquiry. Foxton also wanted an inquiry 'to clear Dr. Roth's good name'. Very little came of either request.

PREMIER BYRNES 1898

Jack was increasingly disillusioned and tired of politicking. Rutledge tried without success to get his Workers' Compensation Bill through. Bill Hamilton was having the same problem with his Shearers' Accommodation Act. No one seemed to be getting anywhere. Normanton and Cloncurry were still to be connected by a rail link but that, too, was thwarted by inter-faction bickering. Jack's old opponent from *The Worker*, Editor Higgs, entered the Legislative Assembly after a stint in local government and moved on again to be Treasurer in a Federal Labor Government. Another recruit from local government was John McMaster. At a subsequent election he lost by 'about 27 votes' and a rumour circulated that he had committed suicide at the result. McMaster, with a dry sense of humour, disregarded the report saying, 'I knew it was a rumour as soon as I heard it'. He not only survived the non-suicide but lived on until well into his nineties.

For some time also, Parliament was running on anything but oiled wheels. In 1897 Nelson and T.J.Byrnes, a Griffith protegee who entered Parliament as the Member for Cairns, went to Britain to attend the Jubilee Celebrations. Byrnes was a young man, born in Brisbane, eighth child in a Catholic family of eleven. When the family moved to Bowen, Byrnes was educated at the local State School and performed with such excellence that he won scholarships to the Brisbane Grammar School. With more scholarships he went on to Melbourne University from which he graduated in Arts and Law. He was considered the shining example for all Queensland children to emulate.

Soon after the Jubilee, Sir Arthur Palmer (who gave his name to the river and the gold field) died and Nelson took his place as President of the Legislative Council. Tozer was Acting Premier in Nelson's absence in Britain but had been appointed Agent General the month before Palmer died and was hence disqualified from standing for the premiership. Byrnes became the new Premier on April 23rd 1898. In late August he attended a Premiers' Conference in Sydney and caught a cold. The cold hung on and was complicated by a bout of measles followed closely by pneumonia. On September 27th, two months before his 38th birthday, his illness, augmented by a cardiac complaint, claimed his life. His death was mourned all over the nation and a fund, the Byrnes Memorial Fund, was set up to help other children in needy circumstances obtain a good education. The T.J.Byrnes medal for outstanding scholarship is still awarded.[113]

THE COOKTOWN CYCLONE OF 1899

Dr.Jack was called back to Cooktown in March 1899 to view the utter devastation in the wake of the cyclone that destroyed the pearling fleet sheltering in Bathurst and Princess Charlotte Bays. More than a hundred vessels, the larger schooners and their fleet of attendant cutters, were lost as well as over 300 lives. The dead included some heroic Aborigines who tried to battle the fifty foot storm surge to rescue the pearlers. Some of the schooners' captains had their wives on board, one with a small baby who, though swept from its mother's arms, together

113. Rosemary Gill, *Thomas Joseph Byrnes* (*Queensland Political Portraits* Murphy and Joyce eds), U. Q. P., 1978 p191

with its mother survived the hurricane's fury. A Darnley Island woman, Muara, whose young husband was killed, swam almost a mile in the darkness and in turbulent seas supporting two injured and exhausted white men to safety. There were many heroes, male and female and of all skin tones. Wreckage from the boats was found a quarter of a mile inland from the beach.

With few and very rudimentary meteorological outposts, Clement Wragge in Brisbane reported a disturbance between the Louisades and New Caledonia but considered there was no danger to the Queensland coast at the time. He later named the cyclone 'Mahina' and alerted the ships along the far northern coast. Unfortunately, it was too late. The damage was already done. Jack was quoted in the Sydney *Bulletin* on his report on the loss of life and of the devastation. 'Birds and mammals lay dead on the land and the beach was wreathed with lines of strange fish, sea animals and marine growths'.

As the result of Dr. Jack's report and as a reward for their courageous assistance, the Queensland Government sent Dr. Roth with a supply of red shirts and bright dresses, tomahawks, knives, a ton of flour, 'two gross of pipes' and a hundred and seventy-three and a half pounds of tobacco to be distributed among the Aborigines along the 'fatal coast'.

PREMIERS DICKSON, DAWSON AND PHILP. 1898

James Dickson took over the Premiership after Byrnes's untimely death in the session that saw Philp's Mining Act made law. Bill Hamilton, ex-shearer and 'conspirator', was then President of the Legislative Council. The Federation Enabling Act was passed by the Upper House and a subsequent act decided Queensland's entry to Federation. There were still a few doubters who thought the young colony might not yet be in a sufficiently strong position to bargain for her rights. Some Separationists had in mind the case in America, where more states were formed after the original thirteen united to form the United States of America. When the referendum was held the 'Yes' case won the day.

In the second session in September, Andrew Dawson confronted Dickson over an amendment to the Railways' Standing Bill requiring 'that the House do proceed to the next Order of the day'.[114] Put to the vote, Dickson had only a 'majority of one' so did the honourable thing for those times and resigned. The Governor sent for Dawson who agreed to form a Government. As Premier, he formed a Labour Ministry – Fitzgerald, Turley, Kidston, Billy Browne, Hardacre and Fisher but contrarily the House had second thoughts and refused to pass

114. D. J. Murphy, *William Kidston* (*Queensland Political Portraits* Murphy and Joyce eds) p233

Dawson's contentious amendment. They had the numbers. Six days after he became Queensland's first Labour Premier, Dawson resigned.

Philp, who had been by-passed for the top job over the years, was called to form a Government. This he did successfully until the end of the life of that Parliament and was then re-elected for a further term. Despite all odds he fought fiercely and with determination for Queensland's interests in the Federation. The Federal Government finally stopped the immigration of South Sea Islanders to the coastal farms. This was a major setback for the canefarmers and Philp tried to alleviate their loss with a tariff on imported sugar – grown with cheap labour elsewhere overseas. In another practical attempt to assist farmers, he established the Agricultural Bank in 1901.

Queensland's finances suffered under Federation when, as predicted by the pessimists, who were also realists, they failed to receive the full share of money due to them. Continuing drought, the tick plague and the collapse of overseas markets reduced crucial export income. With Federation accomplished, Queensland had surrendered her right to raise taxes. Philp, who'd been brought up with the Scots stern 'self help' credo wasn't keen on finding money for the welfare benefits the unemployed were demanding. He was more sympathetic though towards the Aborigines stranded helpless between two worlds. Despite the economies he made, he was re-elected for his second term but the new State's finances worsened. Jack Hamilton's did too and it is reported that after a night session at Parliament, he merely retreated to a small room in one of the back corners and made it his headquarters.

DR. JACK LOSES HIS PARLIAMENTARY SEAT AFTER 25 YEARS

Philp decided he had no alternative but to increase Stamp Duties, one of the few ways left to the States to raise finance. This did not affect the 'working man' and his family so much but it did put him offside with his traditional support base. In addition, banks and trade did not approve. Some of his backbenchers, led by Digby Denham who was a later Premier, changed sides and Philp was left with a narrow majority. Not a 'majority of one' but only slightly more definitive, 33 to 31. Following the current Parliamentary code of honour, he resigned.

The Governor sent for the Leader of the Opposition, Billy Browne, the well-liked Croydon miner. Remembering the short life of the Dawson Government, Browne decided not to make the same mistake. He negotiated a coalition of Labour and the dissidents under the leadership of Arthur Morgan. The unkindest cut for Philp was when his ally of many years, Alfred Cowley, withdrew his support and accepted the Speakership in the new coalition. Jack Hamilton was even more upset over the defection. It had been a tradition for all Northerners to stick together come what may, which was possibly why Philp did nothing about Jack taking up residence in Parliament House. In G.C. Bolton's biography of Philp he quotes Jack as saying, 'We are not sore about the Labour Party beating us. They fought us manfully. What we object to is being, like Lazarus, licked by the dogs.' Cowley's defection was particularly painful to Philp, a man who laid great store on loyalty.

The Morgan Government continued until the end of the term and took on many of Philp's policies. Even the Amended Stamp Act that caused Philp's downfall, was passed without undue delay. Women's Franchise and Workers' Compensation failed at the first new attempt but were passed in 1905. Some of the dissidents regretted their decision but Philp lacked the killer instinct to take advantage of their vacillation. Sadly, Billy Browne died. A motion of no confidence was lodged against the coalition government. To get the numbers, Browne's successor had to be elected. Characteristically, Philp would not consider taking advantage of the situation. He held up procedures until Billy Browne's replacement was elected and took his place in the House. Finally, the No Confidence vote was put and Morgan resigned after the count gave him a bare majority. Philp declined the offer to form a Government and Parliament was dissolved.

The ensuing election was a disaster for the Conservatives, a triumph for Labor. Jack Hamilton lost his seat after twenty-five years in Parliament to a Labor candidate, J.H. Hargreaves. Only seventeen Conservatives survived. Morgan was back with the gifted Kidston as Treasurer.

In many ways Jack was pleased to be out of it. Philp returned to the Premiership again in 1907-8 giving place again to Kidston and a series of Labor Governments but Jack never again sought election. He was 63. He'd had enough of public life.

9

RETURN OF THE MINER 1903-1911

Jack visits Herberton tinfields and far northern goldfields.Hargreaves takes up Jack's fight with Roth.W.R.O.Hill suggests Jack instruct riflemen at Enoggera.Qld Defence Forces absorbed into Commonwealth Forces.Japanese sink Russian fleet.Russian scare abates. Fire at Brilliant mine.Gold at Batavia River and Plutoville. Natural gas at Roma.Cyclone at Cooktown. Earth tremors at T.I. Adult Suffrage (Qld) and PostalVoting (Qld).Old Age Pension (Qld).Burns/Johnston fight.Halley's comet. Floods. T.J.Ryan tries to divide Qld into 3 states. Yongala wrecked.

JACK RETURNS TO EBAGOOLA AND THE HAMILTON KING MINE

Now relieved of his Parliamentary duties for the first time in twenty-five years, Jack was determined to harbor no regrets and to make the best of what years were left to him. There was nothing to take him back to the Palmer. Its busy days, too, were over. It held only ghosts. Rising water drowned the Queen and the claim was abandoned. Efforts were made by a new company to resurrect it in 1901 and 1903 but they failed. The Queen had an unfortunate history since Jack had to finance it through the Queensland National Bank in 1878. It was the deepest mine on the field and being situated on a low-lying flat while other mines were mostly in the ridges, it was all the more susceptible to water intrusion. However in the years from 1877 to 1909 the Queen averaged a little over one and a half ounces to the ton of ore crushed. That gave Jack no little satisfaction. He knew a good mine when he saw it.

The Hodgkinson was all but dead as well but Jack had no desire to return there. His old sparring partner Warden Mowbray died at the end of 1898 in Gympie.[115] At the time of his death he was only forty-eight years old. He was Gympie's Police Magistrate and Warden. A troop of mounted police led his funeral cortege. Veteran Warden, Philip Sellheim, was one of the many who attended. Through Jack's numerous Gympie friends he kept up with news of mining and social activities. The town was jubilant at the record gold production for 1903, over 176,000 ounces of top quality gold.[116] Miners were predicting a rosy future. Jack thought about making Gympie his home again but the call of the Far North was more magnetic. He booked a seat to Cairns on the steamer *Jumna*, took the steam train up Robbs' route to Kuranda and from there went by coach to Herberton.

The township had grown considerably since he'd been on the committee elected to decide the most practical rail route to the coast. The tin mines were flourishing and world demand for the metal high. He accepted the loan of a horse from a friend and took the opportunity to inspect the tin fields. Stannary Hills was booming with a smelter installed at Rocky Bluff on the Walsh River. Optimism ran high. Jack would have liked to visit the copper smelter at Mt. Molloy but time was short, the steamer waited for no man and he was in a hurry to get to the goldfield that bore his name. He wanted to find his mate, John Dickie, but he knew that would not be easy. Dickie wasn't a man to stay in any one place for long.

The steamer called at Port Stewart to land loading for the mines and to allow Jack to disembark. He made his way to Ebagoola and was disappointed, though not surprised, that Dickie was not there. He was prospecting along the Alice River on the new Philp Field. On this trip, Jack had the time to wait and spent it being shown over the workings including Dickie's reward mine, the one that also bore his name, the Hamilton King. The field was reasonably active. Jack was told there were over three hundred miners at work, all making enough to keep them going. Though not in Gympie's class, the mines had produced five thousand ounces for the year.

Two years before Jack's visit, a cyanide plant and a battery to treat the ore were installed at Yarraden some twelve miles from Ebagoolah. Before this, quartz had to be transported by cart to the ten-head Ebagoolah mill, a laborious and costly undertaking. The drought of the nineties affected the mines severely. There wasn't enough water to process the ore. The crushed ore was dry-blown or else stashed in readiness for a break in the weather that would bring enough water to wash it.

115. *Gympie Times*, December 20th 1898
116. *Gympie in its Cradle Days*, Gympie & District Historical Society, 1917 and 1985 p9

Crushing costs were around thirty shillings a ton and carting charges from ten to thirty shillings per ton as well, so that miners couldn't work a mine unless they could count on at least one and a half ounces to the ton. Battery gold, according to Warden Lee-Bryce was worth two pounds nineteen shillings an ounce. Silver was found in conjunction with the gold at Ebagoolah but was not considered a commercial proposition.[117]

Partly assisted by Jack's grant, backing from other friends and a little Government help, John Dickie was an admirable successor to Mulligan in discovering fields but his fascination with exploration won out over the daily grind of working the mines. As was his way, he disposed of his Ebagoolah interests early to finance his prospecting forays elsewhere. Instead of the usual thousand pound reward to the finder of a goldfield, the economising Government paid him only half that, using as an excuse the limited number of miners benefitting – under a thousand – and the low value of the gold, two pounds six shillings an ounce.

On the Hamilton, Dickie tried first for alluvial, it being the simplest to work, but found his best gold in reefs. One miner, Nicholls, left the field with a thousand ounces of gold found in a place he appropriately named Nuggetty Gully. The largest nugget he and his party found weighed almost a hundred ounces. Most of the reefs ran north to south, almost vertical and over three feet wide. Gold was found very close to the surface. Despite the dry times, the return in 1900 was close to 13,000 ounces. The Caledonia P.C. (Prospecting Claim) was the richest mine producing seven ounces to the ton over a crushing of just under forty tons. The Hit and Miss didn't miss too much either. Its tally was 233 ounces from forty-six tons of ore. The Hamilton King wasn't such a spectacular producer but did well enough with very good gold near the surface in its early days.

Unable to contact Dickie, Jack took a coach back to the coast to catch another steamer coming down from Thursday Island. He regretted that he missed another of his old mining friends, Willie Baird. Willie was dead, speared while he and two others were digging a trench at the Bairdsville workings. Jack thought back to the spear wounds he'd successfully treated on the Palmer and wondered futilely if his presence at Bairdsville at the time would have made any difference.

On the boat his thoughts had ample time to wander back to the old Palmer days and the speared miner Paddy Shanahan. In his valise he had the booklet *With the Cape York Prospecting Party* by the other Shanahan. This one, M.W. Shanahan, had stood against Jack for Cook in the elections of 1893. Like Jack, he too, had a

117. *Hamilton Gold Field,* Queensland Mines Dept.

fascination for the furthest North. Jack found the page where Shanahan wrote that the part of the Peninsula near the tip was not 'likely-looking' gold country yet it was given 'consideration over and above anything the geological features of the district would elsewhere warrant.' Jack could understand the allure of the northernmost part of the Peninsula. Gold had been found on the islands to the north, Horn and Hammond, and Surveyor Embley found gold on the beach at Cook's Possession Island. Later, a small strike was made at Peak Head on the mainland directly opposite Possession Island. It'd be an ideal place for a young man to fossick over.

Jack permitted himself a slight smile at Shanahan's farcical picture of him poring over the secret maps that would unlock for him the secret to the Deadman's Gold. He agreed with the more serious Shanahan about the possibility of a gold strike north of the Mein Telegraph Station, and put the book back in his valise. He did think there was a rich field waiting to be discovered – somewhere – but for the present he was heading for Brisbane and then back up to Gympie. The boat called at Cooktown and Jack wondered how the people there were succeeding with their desire to erect a monument to Captain Cook. He'd taken the proposition for Government assistance for the project to Parliament without success. Tozer, when Home Secretary, visited Cooktown where the residents presented him with a list of thirty-seven requests including one for financial help with the monument. This was in the recession of the nineties. Tozer was unsympathetic and rather blunt. 'The position is this: down in Brisbane we have deputations of unemployed asking us for bread; now I come up here and you have asked me for a stone.' As any visitor to Cooktown's foreshore can see, they did get their monument – eventually.

GYMPIE YIELDS A RECORD 179,369 OUNCES IN 1903

Jack hadn't spent much time in Gympie since the floods of 1890 and 1893. The latter, like the Brisbane flood, was the greatest of them all. Compared to it, the inundation of the seventies was a non-event. In the Big Flood almost all the mines flooded and some, because the pressurised, trapped air sought an outlet, blew up.[118] The Raggedy at Monkland was the first to go, followed by Smithfield 2 and 3 which erupted in a terrible explosion with a waterspout a hundred feet high.

Will McNutt told him Skyrings lost more red cedar in the flood. Some was found so far away that the cost of recovery almost exceeded its expected value. It was left where it lay. Will McNutt was married to a Skyring and their home Walla

118. *Gympie in its Cradle Days*, Gympie & District Historical Society, 1917 and 1985 p67

Brindi on Inglewood Hill was open house to Jack whenever he visited the gold town. McNutt, with his experience at one of the richer mines, Monkland 7 and 8, was able to give Jack some sound advice on local mining investments. Production from the Gympie mines rose at the turn of the century mainly due to upgraded machinery and improved mining practices. Mining syndicates were grateful for capital investment in their projects and Jack was pleased with the modest return on his shares. To be back in the industry, even as a very minor but deeply-interested shareholder, gave Jack sincere pleasure as he and Will discussed mining peculiarities and strategies over their customary Sunday dinners at Walla Brindi.[119]

The year Jack left Parliament, 1903, saw Gympiest highest annual gold yield which was a production of 179,369 ounces. With his renewed interest Jack took the record amount as a good omen but unfortunately his optimism wasn't borne out and production slowly declined.

It wasn't long before Hargreaves, the Labor man who defeated Jack in Cook, ran foul of Dr. Roth. He, as Jack had done, complained to Parliament of his constituents being vilified by the Doctor. But this time, many of the Members who had denounced Jack for taking Roth to task, changed sides. The tide had turned. Jack's detractors were now calling for their own inquiries, if not to clear Dr. Roth's name, then at least to clear those of their slandered electors.

Joe Lesina, getting up in the House to defend himself against a Roth attack, declared he 'had always a kindly feeling toward Dr. Roth and repeatedly got up in the House and defended him when he was attacked by the 'late' member for Cook'. He 'had always made a point of standing up for Dr. Roth but… 'The number of charges against Roth were not only growing in number but were getting more serious.' Even the *Daily Mail* published a charge that the Doctor was selling ethnological specimens on his own account. Lesina, stung by Roth's accusations against him personally, interjected during a speech by one of Roth's few remaining supporters, 'If he is guilty of selling public property he ought to be in Boggo Road Gaol!'

Another country Member, Forsayth, who had previously taken Roth's part wholeheartedly in his attempt to stop abuses in the bêche de mer industry, now took the opposing view. He would support the Doctor 'if he could prove his innocence' otherwise he wanted Roth to retract his personal attack on certain bêche de mer boat owners and himself and apologise.

---

119. Conversation with writer's mother, Sadie Waddell, daughter of Will and Tilly McNutt 1996

After innumerable Parliamentary sessions of being the villain, Jack was able to harbour a touch of self-righteous smugness. The irony amused him.

Once settled in his Brisbane lodgings he met an old friend from the Palmer, Willie Hill. W.R.O. Hill was on the Palmer from 1876 to 1878 relieving Warden Coward at Byerstown and later, for a short period, he was the first Warden at the Hodgkinson. Shortly after he arrived in Queensland he took a position with the Native Mounted Police. His brother Cecil, an Acting Sub-inspector in the Native police, died in the service, speared through the heart at the age of twenty-one. Leaving the Police Service, Willie Hill, having also acquired pastoral experience of sheep and cattle runs, headed north to the goldfields. His first assignment was as assistant to Warden Charters at Cape River Goldfield.[120]

THE QUEENSLAND LIGHT HORSE AND JACK HAMILTON THE MARKSMAN AND BOXER

At the time, he was in Brisbane working on his memoirs in which Jack merited an honourable mention. Hill was also something of an athlete, specialising in high jump and pole vault. It was his opinion that Jack's talents should not be wasted. He called on Archibald Meston, no special friend of Jack's but who was helping Hill get his memoirs into shape, to back him up. Especially they considered, Jack should do something with his gift as a sharpshooter. The Defence Force needed instructors. Jack's skills would be invaluable. The idea interested Jack. He'd previously enjoyed his time helping the Cadet Corps in Gympie when his time there coincided with musketry practice and he had been impressed with the ability of some of the cadets. In return, they appreciated his knowledge and skill when he demonstrated the finer points of marksmanship. It didn't take much persuasion for him to accept an instructor's role at the Enoggera army camp.

Queensland was always keen on Defence. Governor Bowen approved the Queensland Mounted Infantry early in 1860. The volunteers provided their own mounts and uniforms and a generous Government found the rifles and ammunition to equip them. Ipswich and Port Curtis formed troops but these were abandoned and the Brisbane based corps changed its name to the Queensland Light Horse, a troop of only some thirty horsemen who, the *Courier* said, were 'most irregular troops in an unorthodox army' Brisbane Grammar School formed a Cadet Corps and other schools followed their lead. Boys of twelve could join the Junior Cadets and continue through their Senior body until they were of an age to join the adults.

120. W. R. O. Hill, *Forty Five Years' Experiences in North Queensland,* Pole Brisbane, 1907.

The Defence Force, all amateurs, strengthened during the Russian scare and when Lt.Colonel French took over command he established a partly-funded Militia. He thoughtfully instituted a fund to provide for the families of men killed in action. Mounted Rifles were established as far north as Charters Towers in the eighties and the mounted men soon gained a fine reputation. They helped keep the peace in the Shearers' Strike gaining experience, discipline and their emu feather plumes. Queensland, New South Wales and Victoria offered troops for the Boer War. The Imperial War Office wasn't over-enthusiastic about accepting their offer, considering the 'Colonials' as 'second class' and undisciplined but Cabinet was eager to cement the Motherland/Colony relationship. They persisted and the War Office gave in. The Colonials, understanding the Boers' commando tactics and bushmanship, outperformed the more formally trained Imperial forces used to wars conducted by the rule book.

Colonel Finn was the last commandant of the Queensland Defence Forces in 1900. After that, the Commonwealth took over the responsibility of defence but Queensland supplied enviable experience and training to the Commonwealth Force. Queensland Generals outnumbered those from any other state and in a Navy List of 1904, 66 of the 135 officers came from the Queensland Marine Defence Force.

It was to a proud defence force at Enoggera that Jack offered his services. Recruits were taken from the 18 to 35 years age group. They had to be at least five feet six inches tall and have an expanded chest measurement of 34 inches. Early rifles were the Martini Henri but their supply was soon upgraded with Lee Enfield 303s with a deadly eighteen inch bayonet. Training was two hours daily for a minimum of two days a week for each group. Sunday evenings were often used for training as well. The Sunday training did not endear Jack to his old friend Father Matthew Horan when he visited the Gympie Cadets.

The Commonwealth Defence Act provided for a volunteer army but also gave the Government power to conscript men from 18-60 years for military duty in time of war on Commonwealth territory. Jack continued to enjoy the time he spent at the rifle range. Some of the soldiers had heard of his reputation with rifle and revolver and his fame spread. Even though he was now past the top limit in age for conscripts his eyesight was still good and his reflexes keen. He took a pride in more than holding his own with the young men at the rifle range. He enjoyed the company and the warm aura of cameraderie it engendered. It took him back in memory to his goldfield days. After years of sitting on Parliamentary benches listening to often tedious speechifying, it was a welcome relief.

Queensland, as McIlwraith predicted, failed to do well financially out of Federation. The state lost its right to impose the greater part of its own revenue-raising charges while the returns from the Commonwealth were neither as generous as promised nor even on an equal pro rata share with the more populous states.

There were over 10,000 miles of Telegraph line in Queensland built on borrowed money. Little funding was coming in to make the repayments. The Commonwealth had agreed to compensate when they took over Post and Telecommunications but that promise had evidently slipped their minds. Jack did not envy Morgan's Treasurer, Kidston, his job of trying to balance the State budget. It seemed an impossible task. To add to the gloom, came news of a fire at the Brilliant Mine in Charters Towers which took the lives of seven miners.[121] It seemed as if Macrossan's work on mine safety still hadn't been enough.

But in some sectors things were looking brighter. The Japanese fleet made a surprise raid in Manchuria on ships of the Russian Pacific Squadron at Port Arthur. The garrison at Port Arthur surrendered after some months of bravely defnding their base. Hostilities continued until the Japanese sank almost all that was left of the Russian fleet. By September 1905, the war was over. For the moment the Russian scare dissipated and Cooktown rejoiced.

On his visits to Enoggera, Jack found his boxing skills were also welcome as the young soldiers sparred in their keep-fit sessions. That he packed a powerful punch many of his sparring partners could verify. He was still particularly agile, light on his feet and with relexes as quick as their own. Years before, Jem Mace, the Swafham Gypsy and one of the last of the great bare-knuckle fighters saw Jack in action on a visit to Australia and remarked that he was one of the best natural light-heavyweight fighters the country had produced. Mace, like Jack, was lightly built by heavyweight standards and was a quick and classy fighter. He travelled widely to the colonies in South Africa, Australia and New Zealand before going on to America to promote the 'science' of boxing.[122]

Neither had Jack forgotten his fencing skills though swordfighting wasn't one of Brisbane's more popular leisure activities. Like the story of the Rockhampton duel there is another, but with evidence to back it up, of Jack fencing blindfolded at the Brisbane Gym in the early 1890s against a 'champion who had the use of his eyes'. Jack was by no means disgraced. The encounter added to his mystique when

121. C. A. Bernays, *Queensland Politics During Sixty Years 1859-1919,* Govt. Printer, Brisbane, p375
122. *Australian Encyclopedia Vol. 4 1987,* Fairfax Syme and Weldon Broadway, N.S.W., p420

it was reported that he parried every thrust without injury to himself and managed to score several times by 'pinking' his opponent with the buttoned foil.

Jack's trips to Gympie, made now by train rather than on horseback or by coach, always included a meeting with his old friend, the Roman Catholic Dean, Matthew Horan. They had much in common. Strong men in a world of the stout-hearted, they were both chivalrous and protective of anyone needing a helping hand, particularly the needy, the 'gentle' sex, the old and the very young. In her late nineties, Willie McNutt's youngest daughter, Sadie, remembered Dr. Jack with evident pleasure as the tall, handsome gentleman who gave her a much-cherished doll and who always had time to speak with her when he came to dinner as if she, too, were a grown-up and his equal. His sincerity and lack of pretension endeared him to the young and old alike.

JACK CONTINUES TO FOSTER MINING EXPLORATION

While Jack invested what money he could in Gympie's mines, he also grub-staked lone prospectors still searching for that elusive, far northern pocket of gold. When Jack had money he was generous and shared it around. Some of his gold dividends reappeared as gold and silver medallions for which the Cadet riflemen competed and some was absorbed in Father Horan's pet charities.[123] Many of his friends practised the same doctrine. Friendship was a bond not taken lightly.

Jack was always pleased to get news of those northermost goldseekers, James Dick and John Dickie. James Dick was most methodical in his ways. He kept records and maps of his trips and meticulously wrote up his mining reports. His friend John Dickie was apt to neglect the paperwork in favour of the excitement of prospecting, much to the disappointment of those who came behind him. Dick, storekeeper as well as miner, was a member of the Cooktown Municipal Council and an ardent supporter of a rail line to parallel the Telegraph Line through to Somerset. He unsuccessfully contested a seat in the Legislative Assembly in 1906[124] but was quite happy to return to his freer life in Cooktown. Prospecting was his first and lasting love but he also had a strong sense of community. He envisaged agriculture taking over in the far north when gold cut out and had an orchard and a successful pioneer coffee plantation on the Endeavour River to prove his point.

Gold was being found further to the north of Coen. An Aboriginal prospector, Pluto, found payable gold on the Batavia River beyond the Mein Telegraph Station in late 1910. Pluto, originally from the Rockhampton district, came north

123. Conversation with Sadie Waddell 1996

124. R. L. Jack, *Northmost Australia Vol 2,* Robertson Melbourne, 1922 p727

with Ned Earl of Butcher Hill. He accompanied Earl and two others on a trip 'Through York Peninsula' in 1896. Ned Earl, as 'Basalt', wrote an account of the trip in the *Queenslander*. Later Earl was prominent in Cairns local politics but in his younger days he led a more flamboyant life which included the riding of the outlaw horse 'Dargan's Gray' before an admiring crowd at Charters Towers.

At the time he found the gold, Pluto could not, as an Aborigine, hold a Prospecting Claim in his own name. He had a white mate, Anderson, in whose name the claims were registered but it was the real discoverer who gave his name to the goldfield, Plutoville, from which he and his mate extracted over two hundred ounces of gold in nuggets, the largest of which weighed 19 ounces. That wasn't the largest found around Plutoville. One tipped the scales at 73 ounces. The gold was of high quality, comparable to that from the Palmer, and returned three pounds, twelve shillings and six pence an ounce.

Pluto found more gold at Chock-a-Block just to the N.N.E. of Plutoville where his biggest find was a nugget of 40 ounces. In 1913 Warden Power reported nuggets of 98 and 121 ounces from the same area. By this time, the number of miners had dwindled. There were only about twenty left. Most of the gold was found in nugget form, not quite the 'ingots' of Deadman's Gold repute but of the far northern fields it came closest to them.

Less valuable minerals were located in the Peninsula including several coal deposits on the North Shore of the Endeavour and along the Laura River. Jack tried in Parliament to interest the Government in further research but the deposits were too far away from cities and towns to warrant the money being spent. Prospectors also knew of the bauxite cliffs at Weipa but with the isolation and lack of amenities no one thought seriously of exploiting the discovery. The tin mines in the Cooktown hinterland, lying closer to a port, were a different proposition and miners operated successfully there for many years.

The year the Japanese/Russian War ended, Morgan's Government successfully abolished plural voting and brought in full adult suffrage. In Queensland, 100,000 women received the right to vote, although their first opportunity didn't arise until the next election in May 1907. As well as that, women were legally allowed to practise law.

In the bush, nothing much changed. The drought continued with the crippling hardship and heartbreak bad seasons bring. Nature was even more vindicative though in the United States of America where the San Francisco earthquake and raging fires from that city's ruptured gas mains claimed hundreds of lives. Unperturbed by that disaster, the small western town of Roma harnessed its natural gas supply to light the town's streets but, the celebrations were barely over when, twelve days after the lights were turned on, the gas ran out.[125] Two years later, the Roma gas bore put on a spectacular display when, somehow, gas and oil ignited and burnt for two months with flames reaching a height of 80 feet.

QUEENSLAND'S OLD AGE PENSIONS AND THE FOUNDING OF THE AWU

Cyclones hit Cairns and Innisfail causing considerable damage but an isolated 'hurricane' which sounds similar to the narrow, extremely intense windstorms of recent years, almost wrecked Croydon, an inland settlement previously considered safe from cyclonic blows. The following season it was Cooktown's turn for a bad cyclone. The Government ketch *Pilot* was sunk and eight people were drowned including Jacks successor as member for Cook, John Henry Hargreaves. Hargreaves, 'a deservedly popular representative' (Matt Fox) was during parliamentary recess, personally attending to queries posed by the government's lighthouse staff. His grave is in Cooktown cemetery. Damage to property in the township was estimated at £20,000. Even little Thursday Island wasn't safe from Mother Nature's wrath. It experienced three minor earth tremors after Cooktown's cyclone had passed.

There was stormy weather in Queensland's Parliament too. Kidston became Premier after Morgan took over the Presidency of the Legislative Council. Kidston was a reformer and he and Philp had several policies in common. One of these that was popular in wideflung electorates was the institution of postal voting. Kidston, however, had trouble within his own party ranks. Digby Denham, who had earlier deserted Philp to let Labor in, now did another U-turn and defected back to Philp.[126] Kidston resigned after an impasse with the Governor, Lord Chelmsford, over the intransigence of the Upper House and Philp was again Premier. It was to be his last stint as Premier/Treasurer as, in an election called the following year, Kidston and Labor regained power. Philp continued on in Parliament until 1915, relaxed and obviously enjoying himself away from the pressures of the top job and basking in the kudos that went with being the Father of the House. Never one to forget a friend, especially one of his loyal Northern Nine, Philp's friendship with Jack continued throughout Jack's life.

---

125. *From Moreton Bay Courier to the Courier Mail* Portside Editions, Fishermen's Bend, Vic. 1992 p95
126. G. C.Bolton, *Robert Philp*. (*Queensland Parliamentary Portraits*. Murphy and Joyce.), U. Q. P., 1978 p213

Meanwhile in Canberra, where things rarely change, the Federal Parliamentarians voted themselves a 50% pay rise from £400 to £600 per year.

Queensland brought in an Old Age Pension Scheme for the over 65s. Jack made it with a year to spare. He celebrated not having to worry what dividends his gold investments would return by going to Sydney's Rushcutters Bay with a group of friends to watch the well-publicised Tommy Burns/Jack Johnson fight.[127] H.D. McIntosh, the Australian promoter, guaranteed the Canadian Burns the then astronomical sum of £6,000 to defend his title against the U.S. negro, Johnson. A special open-air arena was built and the fans flocked in, some 16,000 of them.

Jack was disappointed in the fight. As well as being six inches taller and twelve pounds heavier, Johnson outclassed Burns in every way. He could easily have knocked Burns down in an early round but contented himself playing cat to Burns's mouse, toying with him and completely controlling each round. Burns fought gamely but was no match for the challenger. The fight was stopped in the fourteenth round and the world had its first black heavyweight champion. Looking through the crowd between rounds, Jack searched hopefully for a face and looked forward, against all reason, to finding Billy Scott, the American champion whom he had knocked out in the early days at Gympie and then nursed back to health. Scott was the man who shouted champagne for the bar when in Sydney he heard of Jack's election as Member for Gympie in 1878. Finding one face in a surging crowd of 16,000 was a bit too much to ask, especially after a time lapse of thirty years. Jack didn't even know if Scott were still in Australia but he've had liked to have seen him – for old time's sake.

Compulsory Military Service was introduced in 1909 and the number of military trainees at Enoggera increased. Jack's time was taken up almost entirely by his duties there. Edward V11 awarded battle honours to the 2nd Light Horse for their service in South Africa and the banners were proudly displayed with the 1899-1902 battle record. It was a real morale booster.

Halley's Comet entered the night skies and with it came severe flooding for which it received the blame. Cairns and Innisfail were again inundated, but the drought was well and truly broken out in the bush. Good seasons continued for several years and the tribulations of the long drought were almost forgotten. The first sod was turned on Rockhampton's steam-powered tramway, the only provincial tram service in the nation. Once completed it operated for 30 years providing easy transport for central Queenslanders.

127. P. Arnold, *illustrated Encyclopedia of World Boxing,* Golden Press, Silverton N.S.W., 1989 p34-35

William McCormack and another Union organiser, Ted (E.G.) Theodore formed the Amalgamated Workers Union, the A.W.U. which was to become a formidable force in the future. Tied up in their future too was the discovery of coal at Mt. Mulligan by Bill Harris and Gibbons. This lease was bought by the Chillagoe Company in 1910, the year Andrew Fisher formed Australia's first Federal Labor Government. Back in Queensland's Legislative Assembly,T.J.Ryan who was later to become, like McCormack and Theodore, a Queensland Premier, moved that 'In the opinion of this House the time hads arrived when Queensland should be divided into three states, and when Central and Northern Queensland should each be granted a consitution subject to the Constitution Act of the Commonwealth of Australia.'[128]

His motion hardly made a ripple as it disappeared in the depthless political ocean almost without a trace. Unlike a ship which did cause a great deal of ripples, though for some time it left almost no trace. The *Yongala* from the Adelaide Steamship Company's line foundered in a gale off Cape Bowling Green, south of Townsville. There were no survivors of the 120 crew and passengers. Captain Tutty of the *Alert* saw wreckage which aroused his suspicions, a 16 foot oar followed by cases of kerosene floating at sea. Searching the area, other articles were found which suggested the ill-fated boat was the *Yongala.* The clincher was the discovery of a door with ornate glass panels which was recognised as one of the double doors that gave entry to the music room. The etching contained the word *Festina.* The Adelaide Steamship's motto was *Festina Lente.* Hasten slowly. The *Lente* was etched into the second half-door. The wreck itself lay undiscovered until the days of World War 11.

128. C. A. Bernays, *Queensland Politics During Sixty Years 1859-1919,* Govt. Printer, Brisbane, p532

10

THE DEATH OF THE ANZACs AND THE DEATH OF NOBLE DR. JACK 1915-1916

Mining slump Charters Towers. Titanic sinks. Australian Flying Corps founded. Tramwaymen's Strike. Labor Party's influence gaining strength. Foundation stone laid for Federal Parliament at Canberra. War declared. Rabaul occupied by Australian forces. Queensland women legally allowed to stand for Parliament. Golden Casket inaugurated.

UNION STRIKES, STATE HIGH SCHOOLS AND REGISTRATION OF NURSES. 1912

Then came 1912, a year of mixed fortunes. In Charters Towers, the legendary Daydawn Mine and the Wyndham ceased production and closed. On the other side of the world the *Titanic* sank. On the bright side, the Minister for Defence announced the formation of Australia's airforce, the Australian Flying Corps. With the country's proud record in aviation, there was no shortage of recruits to be trained later at Point Cook. Burley Griffin won the international competition to design the new Federal capital.

In Queensland political events were also mixed. Jack the Hatter's son, Hugh D. Macrossan, a barrister, came into the Nineteenth Parliament.[129] Trade Unionism was gaining an ascendancy among workers and Brisbane tramwaymen went on strike over the suspension of workers charged with wearing union badges after

129. C. A. Bernays, *Queensland Politics During Sixty Years 1859-1919*, Govt. Printer Brisbane, p181

such radical action had been banned. Forty-three unions went out in sympathy and Brisbane ground to a screaming halt. Taken to the High Court, the case was decided in favour of the Unionists. The strike lasted for almost two months and the street violence was often initiated by the Police Force there to reinforce the peace. Fortunately for country Queensland, there was little support one way or the other for either side and life and commerce flowed on. It seemed a small thing to Jack. Whether one wore or did not wear a badge would hardly affect one's working capabilities he reasoned, but Philp, a judicious man in most circumstances, took the wearing of the badges almost as a sign of high treason and impending revolution.

Philp was losing touch with his worker roots.[130] The mates of the young Bob Philp of twenty years before were replaced by a generation of workers more interested in unionism than in individualism. The Labor Party's influence was rapidly spreading.

State High Schools were opening in diverse areas of country Queensland and a registration system for nurses was introduced into Queensland hospitals. It was so successful that other states adopted it as a model for their own organisations.

Once again Cooktown was damaged by cyclone and Cairns and Innisfail saw more record flooding. Jack was anxious to see the extent of the havoc but with finance limited to his modest age pension and his commitments at Enoggera and Gympie, he had little choice but to wait patiently for the reports to filter through to Brisbane along the grapevine and in the Press. For some time the grapevine had predicted the closure of the Mt. Molloy smelters and 1913 proved the forecasters accurate. Individual miners were still making a bare living in the northern Peninsula, finding yet another nuggetty gully as the last one's supply became exhausted but no one was making a fortune. The work entailed was hard and the isolation at times overpowering. The Deadman's ingots were still safely hidden.

It was a great year for the Commonwealth when the wife of the Governor General, Lady Denham, christened the site of the new Capital Territory, Canberra. Her husband, assisted by Prime Minister Andrew Fisher and the larger-than-life Home Affairs Minister, King O'Malley, laid the foundation stone for the new building. A couple of months later, Labor lost its majority in an election and Joseph Cook replaced Andrew Fisher as P.M., his third occasion in that office.

130. G. C. Bolton, *Robert Philp* (*Queensland Parliamentary Portraits.* Murphy and Joyce) U. Q. P., 1978 p217

QUEENSLAND'S SPIRITED BUT TRAGIC DEFENCE OF EMPIRE. 1914-15

When 1914 dawned, the drums of war were already thrumming in Europe and coincidentally or by intent, General Sir Ian Hamilton[131] began an eleven weeks tour of the Empire's Australian troops. He arrived at Enoggera in an abnormally hot and humid February but was most impressed with what he saw. The Light Horse really captured his admiration. 'They are the most formidable body of troops who would shape very rapidly under service conditions'. He was right and it was to be proved even sooner than most war-mongers predicted.

Prime Minister Cook promised 20,000 men[132] including a Light Horse Brigade of 2,000 mounted soldiers. Men flocked to join up from all over the nation. Adventure was in their blood and the defence of the Empire the biggest adventure available to them at the time. Light Horse enlistments exceeded those of the infantry and other services. In July, the Archduke Franz Ferdinand of Austria was assassinated at Sarajevo. The war drums pounded out an intoxicating tattoo. Germany occupied Luxembourg, invaded Belgium and declared war on France, all in a matter of days.

Britain declared war on Germany on August 4th but it was Australians who fired the first shot on the Allies' behalf. A German merchant ship, the *Pfalz*,[133] was moving out of Port Phillip Bay when news of Britain's Declaration of War was telegraphed through. That meant Australia, too, was at war with Germany and the *Pfalz* at once became an escaping enemy vessel. The battery at Fort Nepean didn't hesitate. A warning shot was fired across the *Pfalz*'s bows. Taken by surprise, the Captain immediately surrendered without need for a second shot and returned his ship to the berth it had just vacated. Unexpectedly, the repercussions soon reached Chillagoe. Most of the smelter's financial backing was German-based and, with the start of hostilities, their assets were frozen.

The plot thickened and the action increased. Within a fortnight, an Australian Military and Naval Expeditionary Force of 1500 left Sydney for that bone of contention to Queenslanders, Kaiser Wilhelm Land or German New Guinea. Rabaul was occupied and the German Governor surrendered to the victorious invaders. Meanwhile, the *H.M.A.S. Melbourne* captured an enemy radio base on New Britain which had been trying to alert the Fatherland of the unaccustomed busyness in the sleepy Pacific. Sadly, the action was not accomplished without loss. Six lives were sacrificed, the first Australian casualties of the war.[134]

---

131. Starr and Sweeney, *Forward*, U. Q. P., Brisbane, 1989 pp55-56
132. *Brisbane Courier*, August 3rd 1914
133. *Brisbane Courier*, August 5th 1914

Another casualty off New Britain was one of Australia's few submarines, the *AE1.*[135] Some weeks later she was avenged when the *H.M.A.S.Sydney* disabled the German *Emden*[136] off the Cocos Islands. The crew was captured and the vessel sunk.

Jack was of two minds about the dispatch of so many high-spirited young soldiers to the war zones. His adventurous temperament longed to go with them yet he had a niggling doubt it wasn't his kind of war. It wasn't like Juarez struggling for democracy against imperialistic power. What did the assassination of an Austrian Archduke have to do with the young men of the 2nd Light Horse and with Australia? It seemed more like a power struggle than a fight to uphold justice but once they were occupied by the enemy, the small non-combatant countries had to be liberated at all cost.

Perhaps it was fortunate, as he joined the crowds at Pinkenba to farewell the troops with their distinctive slouch hats and 'kangaroo feathers' that neither he nor the others saying farewell could foresee the bungling incompetence of some of the British staff officers which was to sacrifice the lives of thousands of the young men they were so blithely seeing off their war. Sailing in a convoy of forty large transports escorted by British and Japanese warships, the men were destined for a freezing training ground on Salisbury Plain prior to serving in France. Chauvel, concerned that his Light Horsemen, poorly equipped for the northern climate, were to camp under canvas in an English winter,[137] prevailed on the powers that were to disembark the men in Egypt to train in more temperate conditions.

The senseless carnage of Gallipoli was an unbelievable nightmare yet to be endured. The severity of the losses there shocked those waiting at home and when the casualty lists were made public the heartbreak was universal. Many of the young men whom Jack had assisted in training, had joked and exchanged witticisms with, would never return. One of these was a namesake and a favourite of Jack's. Notification of his death appeared on one of the earliest lists, 'Private John Hamilton, 21. Died of wounds.' Jack hoped the result could justify the overwhelming loss of young life. Though he kept up his work with the Enoggera recruits and the cadets at Gympie, his heart was no longer in it. Nor was his health good. However he considered that he owed it to the courage and the

---

134. *Brisbane Courier,* September 11th 1914
135. *Brisbane Courier,* September 14th 1914
136. *Brisbane Courier,* November 9th 1914
137. Star and Sweeney, *Forward,* U. Q. P., Brisbane, 1989 p57

resourcefulness of the young soldiers to see that they went to battle with the best available training. They deserved at least that much of a sporting chance.

It was a grim year too in the rural areas. Many of the young bushmen were the first to enlist – and to die. At home, their families were trying to cope without them and with yet another crippling drought. All the while they were being exhorted to produce more meat to feed our brave boys and more wool to keep them warm in the chilly trenches. But 1915 was the year in which politically-minded Queensland women advanced one more feminist step. As well as being able to vote for a male candidate, they were now eligible to contest a parliamentary seat in their own right.

THE DECLINE AND DEATH OF DR. JACK. 1916

There was a sad start to 1916 when news gravitated down from the Peninsula that the prospector Pluto was dead.[138] Another link in Jack's golden chain was broken. Some years later, Kitty Pluto carried on the name by ('accidentally') finding gold at Lower Camp, a bit over a mile away from Plutoville when she located a nugget while carting surface wash to the river from Pluto's claim. Jack was now past his three score years and ten. Although he was loath to admit it, he was slowing down. What had always been a pleasure was now more like a chore but he was reluctant to sever completely his tie with Enoggera. He was needed and that was enough for him. His constant headaches were something he'd have to suffer without complaint. It was the least he could do for his young charges.A sympathetic commandant decreased his hours but also saw Jack's need to be 'doing his bit'.

The first Anzac Day was held to commemorate the anniversary of the landing at Anzac Cove. All troops evacuated from that hell on earth were diverted to the muddy fields of France or to the deserts of the Middle East. Prime Minister Billy Hughes talked of conscription to keep up the number of available servicemen and was expelled from the New South Wales Labor Executive.

Jack permitted himself a wry smile, a rare occurrence in those times, when he heard of the move to start a lottery to raise money for the War Effort. The corrupting powers of the sweeps and consultations[139] were conveniently forgotten in the need to finance a war. The name itself was beguiling – the Golden Casket. For a few pence outlaid it promised fortune and the realisation of a lifetime's dream.

---

138. R. L. Jack, *Northmost Australia Vol 2*, Robertson Melbourne, 1922 p735
139. *Brisbane Courier*, December 1916

As the year progressed Jack's health deteriorated. He was confined most of the time to his lodgings in a modest Brisbane boarding house. He was plagued with excruciating headaches that would not go away and his muscle-coordination, up till then barely affected by age, rapidly declined. With his medical knowledge he had grave doubts about his condition but tried to dispel any discouraging thoughts.

Will McNutt and Matt Horan tried to convince Jack to return to Gympie. There was always room for him at Will and Tilly McNutt's home. At the onset of his disability he was persuaded to visit his friends at Gympie for short periods.Young Sadie was devastated to see her hero in such diminished health and determined to follow an elder sister's example and become a nurse. With nursing skills she was confident she'd soon have Dr. Jack back to his usual exuberance. He showed obvious pleasure when taken to inspect the Gympie cadets at training and the lads, who greatly admired his skills, were equally pleased to see him.

In time, the journey to Gympie and back became too tiring and his trips were sacrificed. He resented not being able to do as he would like. No obstacle had previously been too difficult for him to overcome. His helplessness rankled his independent spirit but there was little he could do to alleviate it. Often he wondered about what might have been. What if he had made an early fortune and sailed to Mexico to aid Juarez? Would it have hastened the arrival of democracy in Mexico? What if he'd agreed to Slippery Sam Macalister's suggestion that he take the Gold Commissioner's job on the Palmer? Would that have made a difference? Or if he'd taken the contingent of experienced soldiers to the Maori Wars?

Had he made a mistake when he decided against standing for Cook in early 1876? He thought not. Those extra two years were more usefully spent on the goldfields even if his problems with Warden Mowbray did result in a Select Committee Inquiry in 1879. His name was cleared as he and his friends had known it should be. If he had married would he have had more companionship and someone to care for him now? He dismissed that thought categorically. His friends were all he needed. They'd never let him down.

He still took an interest in Gympie's mines though he no longer owned any shares. Yields were gradually declining since the dramatic resurgence in the first decade of the new century.[140] Even so, 1916 still saw close to 50,000 ounces being produced to bring the field's grand total to just under four million ounces of gold.

140. *Gympie in its Cradle Days*, Gympie & District Historical Society, 1917 and 1985 p10

Friends continually offered help but, always a loner and an active athletic man, he strove valiantly to hang on to his independence until the beginning of October when he was no longer able to cope by himself. Anxious friends took him to the Brisbane Hospital. The diagnosis was not good. Cancer of the spinal cord. His body, which had never before let him down, now refused to obey even the most simple signals his brain tried to transmit to it. He was absolutely helpless. Within days, he lapsed into a coma which prevailed until he died at 10.30 a.m. December 7th 1916.

Jack left no will and extremely few earthly possessions. What he acquired through life he generously shared with others, rather than try to amass an estate for distribution after his death. The slate was clean.

At his funeral there were many eulogies for this well-loved, almost legendary, figure, a knight without fear of reproach, a great gentleman, a gallant enemy depending on viewpoint but Matthew Horan who knew him as well as anyone, spoke for all his friends when he responded, *'I have read the stories of saints and martyrs and the motives which induced them to live such becoming lives. Mr. Hamilton's motives I do not profess to consider. That is the divine prerogative of the great searcher of hearts. Nor do I know what form of Christianity he professed. But I do know that, for the long term of years I knew him, his life has been the scene of good actions towards neighbours based upon that of the Good Samaritan so highly commended by the Great Master'.*[141]

Willie McNutt, himself no longer a fit man, made the trip from Monkland to Brisbane with his wife to attend his friend's funeral but sixteen year old Sadie stayed home at *Walla Brindi*. Her grief was inconsolable though her older sisters tried to comfort her. The gentlemanly Mr. Hamilton was her idol as she grew up in Gympie. A constant visitor to her home, he always had time to talk with her and took pleasure in making her smile. He told exciting tales of life on the Gympie and far northern goldfields but modestly, rarely told of his own part in the action. She took from its high shelf the beautiful doll he'd sent to the city to get for her so many years before. She'd treasure it for ever. Clutching the dearly-loved Lottie for comfort, she cried herself to sleep, heartbroken over the loss of her childhood hero.[142]

141. *Gympie Times*, December 9th 1916

142. Conversation with writer's mother Sadie Waddell nee McNutt 1997.

BIBLIOGRAPHY

NEWSPAPERS & RECORDS

Brisbane Courier.

The Worker, Brisbane.

Courier, Cooktown.

Herald, Cooktown.

Morning Post, Cairns.

Hodgkinson Mining News, Kingsborough.

Gympie Times.

The Sydney Bulletin.

Votes and Proceedings, Qld. Parliament.

Baptismal Records, Victoria.

Shipping Lists 1863/4 Bowen Historical Society.

Records of Queen of the North 2 Mines Dept. Archives.

Pugh's Almanac 1885.

Royal Commission 1881 Mr. Hemmant's Petition. Govt. Printer.

Death Certificate and Public Curator's report on estate of John Hamilton.

BOOKS & PERIODICALS

Ball Lionel C. *Hamilton and Coen Goldfields Brisbane 1901* Brisbane 1901

Bell Peter *Gold Iron Steam* J.C.U. Press 1987

Bennett M.M. *Christison of Lammermoor* London 1927

Bernays C.A. *Queensland Politics during Sixty Years 1859-1919* Govt. Printer Brisbane 1919

Bicknell A.C. *Travel and Adventure in Northern Queensland* Longmans, London 1895

Bolton G *A Thousand Miles Away* Jacarandah Brisbane 1963

Browne Spencer *A Journalist's Memories* Brisbane 1927

Collinson J.W. *Early Days in Cairns* Smith Paterson 1939

Collinson J.W. *Tropic Coasts and Tablelands* Smith Paterson 1941

Collinson J.W. *More About Cairns (3 vols.)* Smith Paterson 1942, 1945, 1946

Cooktown and District Historical Soc. *Greatest of All Cyclones*. n.d.

Corfield W.H. *Reminiscences of Queensland 1962-1899* Pole and Co Brisbane 1920

De Havelland D.W. *Gold and Ghosts Vol. 4* Australian Print Group 1989

Dorothy Jones, *Trinity Phoenix, Cairns Post* 1976 p58

Grabs C. *Gold Black Gold and Intrigue* Angus and Robertson 1982

Frost Cheryl *The Last Explorer* J.C.U. Press 1983

Gympie and District Historical Soc. *Gympie in its Cradle Days* Gympie Times 1917 and 1985

Hill W.R.O. *Forty-Five Years' Experiences in North Queensland* Pole and Co. Brisbane 1907

Jack R.L *Northmost Australia Vols. 1 and 2*. Simpkin 1922

Kennedy K ed *Readings in North Queensland Mining History Vol1* J.C.U. Press 1980

Knowles J.W. *The Cooktown Railway* A.R.H.S. Queensland 1966

Murphy and Joyce eds. *Queensland Political Portraits* U.Q.P. Brisbane 1978

Palmer E. *Early Days in North Queensland* Angus and Robertson 1903

Pike G. *On the Trail of Gold* Watson Ferguson and Co., Brisbane 1998

Robertson Jillian *Lizard Island* Hutchinson n.d.

Shanahan M.W. *With the Cape York Prospecting Party 1896* n.d

Starr and Sweeney *Forward* U.Q.P. Brisbane 1989

Svensen S. *The Shearers' War* U.Q.P. Brisbane 1989

Books by

Lennie Wallace

Bow Waves in the Bull Dust

Bitten by the Bull Bug

Nomads of the 19th Century Queensland Goldfields

The Battlers of Butchers Hill

Cape York Peninsula

From Nanango to Cooktown